# The Demography of Europe

Gerda Neyer · Gunnar Andersson
Hill Kulu · Laura Bernardi
Christoph Bühler
Editors

# The Demography of Europe

Springer

*Editors*
Gerda Neyer
Gunnar Andersson
Department of Sociology
Stockholm University
Stockholm
Sweden

Hill Kulu
Department of Geography
University of Liverpool
Liverpool
UK

Laura Bernardi
Life Course and Inequality Research Centre
University of Lausanne
Lausanne
Switzerland

Christoph Bühler
Institute of Sociology
Leibniz University Hannover
Hannover
Germany

ISBN 978-90-481-8977-9 ISBN 978-90-481-8978-6 (eBook)
DOI 10.1007/978-90-481-8978-6
Springer Dordrecht Heidelberg New York London

Library of Congress Control Number: 2013933245



Printed on acid-free paper

Springer is part of Springer Science+Business Media (www.springer.com)

# Contents

# Chapter 1
# The Demography of Europe: Introduction

**Gerda Neyer, Gunnar Andersson and Hill Kulu**

Over the past three decades Europe has witnessed fundamental changes in its population dynamics and population structure. Fertility has fallen below the replacement level in almost all of the continent's countries, while childbearing behavior and family formation have become more diverse. Life expectancy has risen in Western Europe for both females and males, but has declined for men in some Eastern European countries. Migration from non-European countries to Europe as well as cross-border mobility within Europe have increased substantially. These changes pose major challenges for population research. The variety of their trends and patterns make it difficult to assess the direction, the speed, and the quantum of demographic changes and to draw general conclusions about Europe's demographic pathway. Conventional theoretical assumptions regarding patterns of demographic behavior and demographic development seem to be increasingly unfit to provide convincing and comprehensive explanations of the recent demographic developments. The dynamics of change and the diversity of demographic behavior call for further advances in demographic approaches and methods. This would facilitate the understanding of the complexity and interrelatedness of factors influencing demographic development. In addition, the growing political interest in demographic issues has increased the demands on demographers to provide substantive evidence of the causes and consequences of Europe's population

G. Neyer (✉) · G. Andersson
Department of Sociology, Stockholm University, Universitetsvägen 10B,
106 91 Stockholm, Sweden
e-mail: gerda.neyer@sociology.su.se

G. Andersson
e-mail: gunnar.andersson@sociology.su.se

H. Kulu
Department of Geography, University of Liverpool, Roxby Bldg L69 7ZT Liverpool, UK
e-mail: hill.kulu@liverpool.ac.uk

G. Neyer et al. (eds.), *The Demography of Europe*,
DOI: 10.1007/978-90-481-8978-6_1,

development. This has increased the pressure to expand research towards assessing the impact of social, economic, and political factors on demographic behavior and demographic trends in Europe.

These issues have been a driving force behind much of recent demographic research in Europe. A large body of literature deals with various aspects of European population development and population dynamics, employing a variety of different approaches, different methodologies, and different aims. The articles in this book contribute to this literature. They focus on specific demographic issues relevant to the understanding of the diversity and of the complexity of Europe's demographic development. The chapters span across all the main demographic areas: fertility, family dynamics, mortality, and migration. The aim of the book is twofold: first, to present and discuss some features of population issues which are vital to the demography of Europe, and second, to address issues associated with researching these features. The book also offers examples which illustrate that the demography of Europe, that is, the demographic issues and demographic development of Europe as we know them, cannot be separated from demography as a science, that is, from the theoretical and methodological approaches that underpin demographic research. The contributions also call attention to the need for cross-disciplinary research, for broadening theoretical perspectives and for advancing methodological approaches in order to better understand the complexity of demographic behavior and demographic development in Europe. In what follows we summarize the contributions in this book and embed them in the broader context of demographic development in Europe.

## 1.1 Fertility and Family Formation

For more than a decade, from 1994 to 2005, the total fertility rate (TFR) of Europe hovered around 1.45. Some countries had this low or even lower fertility for several decades; others experienced a drop of their fertility to low (1.7–1.5 TFR) or lowest-low (below 1.5 TFR) levels only in recent years. Only in a few countries has the TFR never fallen to low or lowest low levels. Despite the recent increase in the TFR in all major regions of Europe (Frejka and Sobotka 2008; Goldstein et al. 2009), the persistent below-replacement levels of the TFR have given rise to concerns about the consequences of low fertility for the sustainability of Europe's economic, social, and welfare systems.

In his contribution, *Peter McDonald* notes that Europe's entrenchment in low fertility poses a major challenge for demographic research. He highlights five areas on which demographic research should concentrate in order to improve our ability to predict fertility trends, to enhance the knowledge of the causes of low fertility, and to assess the consequences of it. First, he sees a need to develop better fertility measures based on detailed indicators of childbearing behavior in order to overcome the widely acknowledged weaknesses of standard demographic fertility measures and in particular the TFR. Developing new measures requires the use of longitudinal data and of methods which follow women's childbearing behavior over their life course (see also: Hoem and Mureşan 2011a, b; Hoem et al. 2011a).

Second, McDonald argues for concerted efforts to examine the relationship between institutional factors and individual behavior. He calls for theoretical models and empirical methods which acknowledge institutional change explicitly and which capture cumulative impacts of institutional factors. Third, he feels that the impact of values on childbearing behavior is not yet understood adequately. Prevailing approaches are too simple to explore the diversity of the effects of values on childbearing behavior, to grasp the strengths of values in the decision-making process, and to explain cross-national value differences and their impacts on fertility. The methodological repertoire should be broadened towards mixed-method approaches, that is, towards incorporating qualitative research methods in demography along with the quantitative approach normally in focus.

Fourth, demography suffers from a weakness in theories. Even though the attitude that demography is a science without theory may be exaggerated, the most common theoretical approaches in demography do not seem to grasp the diversity and directions of current fertility and family development in Europe. According to McDonald we need theories which capture change, at the institutional, at the value, and at the individual level. Fifth, demography should open up more towards policy-oriented research. It should focus on estimating the impact of policy packages on demographic behavior, taking into account the range, coherence, and perceptions of policies. There is also a need to further assess the reasons for equal or differential effects of policies on demographic behavior of different groups of the population.

According to McDonald these challenges can only be tackled if demography pushes for theory-based approaches, for cross-national comparative longitudinal (panel) data which allow investigations of the heterogeneity of childbearing decisions, for the advancement of methods which facilitate analyses of such data, and for a more consistent incorporation of qualitative methods in demographic research.

Following Peter McDonald's suggestion to explore the relationships between various aspects of policies and fertility, *Gerda Neyer* discusses the potentials of welfare-state and of policy approaches for fertility research in Europe. The low total fertility rates in Europe during recent decades have spurred suggestions that countries should implement policies in order to raise fertility. While such demands flourish in public discourses, demographers often find no or only small fertility-elevating effects of individual policies. This has led them to turn their attention to the welfare state instead, that is, to national configurations of social and family policies and their effects on fertility. Neyer argues that welfare-state and single-policy approaches complement each other. She points to two ways in which welfare-state oriented research can contribute to a better understanding of the linkages between policies and fertility. First, despite many recent changes of welfare policies, the configurations and main directions of European welfare states have not undergone substantial transformations. This inertia opens up the possibility to assess the cumulative impact of policies and policy packages over time and over a person's life course. Second, the concept of the welfare-state regime as developed by Esping-Andersen interlinks welfare-state policies, the labor market, and the family. The "varieties of capitalism"-approach broadens this further in that it associates different types of welfare states with different labor-market structures and different educational systems. This permits researchers to see fertility behavior within a framework of interrelated institutions. Drawing on comparative work on ultimate childlessness by educational attainment in Sweden and Austria, Neyer demonstrates the usefulness of such an approach. It offers explanations of differentials in cumulative fertility outcomes that cannot be linked immediately to individual policies or single policy packages.

Analyses of individual policies provide insight into their effects on childbearing behavior, which makes it possible to distinguish between differential effects on the fertility behavior of various groups of women and of men. If the welfare-state configuration and the socio-economic context are considered in the analysis and/or in the interpretation of the research results, such analyses are also well suited to assess the potential effect of a policy in other contexts. Neyer gives examples of these features based on research findings from Sweden. Since analyses of individual policies provide a more nuanced picture of the effects of a policy on different social groups than more macro-oriented analyses, they are also useful for policy advice. Researchers can only make use of this potential if detailed data are made available that also reveal the effects of special regulations geared at specific social groups.

The low levels of the TFR in Europe are often attributed to the increase in the mean age at childbearing. Over the past 30 years, women in all European countries have delayed childbearing considerably. In almost half of the member states of the European Union, women's mean age at childbearing is currently 30 or above (European Commission 2011: 29). Demographers give several reasons for the postponement of motherhood: the expansion of the time spent in education, difficulties in the transition from education to work, changes in union-formation and marriage behavior, greater individualization, and growing cleavages between women and men in their attitudes towards parenthood. However, the considerable variation in women's age at first birth and in their mean number of children lead us to question the prevailing assumption that educational, economic, and social factors have a universal, gender-neutral, and uniform effect on postponement (Andersson et al. 2009b; Neyer and Hoem 2008; Van Bavel 2010; Kantorová 2004; Kreyenfeld 2010; Andersson et al. 2009a). This has stressed the need for research which explores how institutional, economic, and cultural circumstances intertwine with the life course and with attitudes of women and men in shaping their childbearing decisions. Such research needs to go beyond the usual analysis of quantitatively measurable life course events; it needs to grasp how people perceive their own circumstances, what meaning they give to their own life course trajectories, their education, their employment, and their personal relationships, and how these influence their decision whether and when to have a child (Bernardi et al. 2008; Keim et al. 2009).

The chapter by *Karl Ulrich Mayer and Eva Schulze* is one of the few studies in demography which uses such a multi-dimensional framework and combines quantitative and qualitative methods to explore the reasons for postponement of first births, with an application to reunified Germany. The past division of Germany into two countries with different educational, economic, gender, and policy regimes makes it particularly well-suited for a study of the impact of educational, economic, and social factors on childbearing decisions. Analyzing the timing of first birth across cohorts, Mayer and Schulze depict the emergence of different patterns of postponement in East and West Germany. Compared to their West German counterparts, women in the former East Germany had their first child earlier, their educational attainment had no marked influence on the timing of their first birth, and they did not sequence marriage and childbearing to the extent that West German women did (see also: Kreyenfeld 2004; Bernardi et al. 2008).

Mayer and Schulze ask whether these different trajectories to motherhood in the two Germanies have left their imprints on the childbearing behavior of East and West German women after unification. Their narrative interviews confirm that they have. They document nicely the different motives and the different behavioral patterns in East and West German women's pathway to motherhood. The different histories of East and West Germany have led to lasting differences in women's perception of motherhood as an integral part of their life, in their expectations for reconciling work and childrearing, in their reliance on a stable partnership as a precondition for childbearing, and, correspondingly, in men's willingness to commit themselves to a family and to take up the responsibility of fatherhood.

The study by Mayer and Schulze demonstrates the gain in insight obtained through a research that combines quantitative life-course analysis with qualitative interviews. It highlights, first, that societal institutions do not only structure life courses, but that they also influence how people assess constraints and possibilities. Second, their study underlines that a society's past has repercussions on current individual decision making and life-course planning. Third, it emphasizes the need to explore family formation and demographic issues of the family from a couple perspective. Fourth, it exemplifies that childbearing decisions do not follow a pattern similar in all societies and also that women and men may consider individual and societal circumstances differently in their plans to form a family. Finally, it stresses the need to uncover the motives, the attitudes, the ambivalences, and the expectations which underlie the decision to whether or when to form a family and to have children.

The persistence of country-specific behavior over time suggests that there exist standardized or normatively regulated patterns of behavior that are particular to each society (Liefbroer and Billari 2010; Billari et al. 2011). Nevertheless, increasing intra-country variation in family and fertility matters indicate that norms become weaker and prevalent patterns of behavior become de-standardized. Explanations that the recent changes in European family and fertility patterns are the consequence of de-standardization figure prominently in demographic research. For example, many of the changes in the demographic behavior of Europeans are regarded as the outcome of increased individualization, of the dissolution of traditional family configurations, of the erosion of gender relationships and the increase in women's autonomy, or of the diffusion of previously uncommon behaviors within the population. In his contribution to this book, *Johannes Huinink* takes issue with the notion of de-standardization. He argues that the prevalent concept of de-standardization is too narrow and uni-directional to capture and explain changing life course patterns. He calls for a more nuanced notion of de-standardization which should distinguish between changes which are irreversible and changes which are transient, and he advocates considering *re*-standardization. Huinink demonstrates the need for a broader concept by looking at the changes in the transition to adulthood across cohorts and across countries. The transition to adulthood is a particularly interesting demographic phase to study in connection with issues of de-standardization of behavior because it comprises several transitions: leaving the parental home and forming one's own household, first union formation, first marriage, first birth, finishing education, taking a first job, and so on. Concentrating on the age structure of demographic events in the transition to adulthood, on the interrelationship and sequencing of these transitions, and on the diversity of demographic, educational, and work-related trajectories at the threshold to adulthood, Huinink shows that the changes in these transitions cannot be captured by a single concept of de-standardization. He finds that some of the changes observed across cohorts were only temporary, and that younger cohorts show a behavioral pattern similar to their parent or even their grandparent cohorts. In some cases, a new form of behavior replaced the previously prevalent form of behavior, so that it is more appropriate to speak of re-standardization than of de-standardization of behavior. Some behavioral patterns have become more diverse over time. But Huinink cautions against terming such changes de-standardization without further consideration. It could well be that these changes are also only transient and that the behavior will either reverse to the previous standardized form of behavior or end in a new form of standardized behavior. In light of these findings, Huinink pleads for the development of concepts of social change which are less concentrated on comparisons of behavioral states, and are built on assumptions about the logic of processes of change. Such an approach would also facilitate an assessment of whether processes of change occur simultaneously in different areas of behavior or whether change in one behavioral pattern lags behind change in another. This could provide insight into the demographic consequences of synchronic or a-synchronic processes of change and shed light on some of the so far unexplained issues of the demographic diversity of family and fertility patterns in Europe.

## 1.2 Longevity and Mortality

The low fertility and the changes in family patterns in Europe have not only attracted attention as such, but they have been increasingly discussed within the broader context of population aging due to the decrease in mortality and the concomitant rise in life expectancy. During the twentieth century European mortality risks have fallen substantially and life expectancy has been rising steadily. Life expectancy has been increasing by two and a half years per decade in countries with the highest recorded life expectancy (Oeppen and Vaupel 2002; European Commission 2009: 19). The European Commission projects that in the member states of the EU life expectancy at birth will continue to increase over the next half century. By 2060 the European Commission reckons that life expectancy at birth in Europe will be 89 years for women and 84.5 years for men (European Commission 2009: 36). According to the United Nations, Europe is the continent with the highest share of elderly (people aged 60 and above) in the world (Rau et al. in this volume). Given the increase in life expectancy, the low fertility rates, and the selective immigration, Europe is likely to retain its rank as the oldest continent for the foreseeable future (Rau et al. in this volume).

The aging of the population is often assumed to have severe repercussions, for example, for Europe's economic growth, its productivity, its care needs, and for the sustainability of its welfare state. Yet, research shows that on average Europeans do not only get older, they also stay healthy longer. *Roland Rau, Magdalena Muszyńska and James W. Vaupel* take this development as their starting point to discuss the development of the various faces of aging, of mortality, and of life expectancy in Europe. They concentrate mainly on four countries: Norway, Denmark, Sweden, and the former East Germany prior to and after unification. The variations among them show that even countries with similar welfare states, with similar gender patterns of employment, or with other similar aspects (such as Norway, Denmark, and Sweden) may have different developments of mortality and life expectancy. Life expectancy in Denmark has not followed the trend of the other Nordic countries: women and men in Denmark have lower survival rates and partly larger inequality in lifespan than women and men in the other countries. Moreover, due to the rapid increase in life expectancy in East Germany after German unification, the life expectancy for Danish women is now below the life expectancy in East Germany. Rau, Muszyńska and Vaupel point out that for all countries it is particularly the increase in survival at ages 65–80 which is remarkable, because improvements in life expectancy above age 65 are normally considered to be difficult to achieve. In 2005, about 50 % of all women and men who reach age 65 can expect to survive to ages 82–84 (men) or 85–88 (women), and 10 % to ages around their mid-90s.

The authors also show that the increase in healthy life expectancy has recently even exceeded the increase in total life expectancy. Even at age 60, women in the four countries studied can expect to live their life in full health for another 17–20 years. Rau, Muszyńska and Vaupel regard this as an opportunity to discuss the re-structuring of the standard patterns of education, employment, and retirement across the life course in order to tackle the social, economic, and welfare consequences of population aging.

The remarkable differences in life expectancy among and within European countries have inspired much research as to the causes of such differences. In his contribution *Jacques Vallin* examines them in detail and tests theoretical approaches to the epidemiological and health transitions which aim to explain the changes in life expectancy and mortality over time. The development of life expectancy in European countries since the 1950s has resulted in marked cleavages between Eastern and Western European countries. Life expectancy in most Eastern European countries now ranges from 65 to 73 years and in Western European countries from 78 to 82. As Vallin shows, this East–West divide and the greater differences in life expectancy in Eastern Europe are due to a convergence of and an increase in life expectancy in the West and to a more heterogeneous development in the East. In some Eastern countries, life expectancy even declined for some years and mortality among (male) adults rose significantly. Among the many factors which may account for such diversity, Vallin investigates geographical, gender, and occupational differences in mortality over the past decades. He concludes that differences in causes of death and differences in the gains from medical and health improvement may account for the regional, gender, and socio-cultural differences in mortality. Yet, neither the current differences nor the diverse developments of mortality patterns in Europe can be adequately understood without taking the historical trends in socio-economic or epidemiological development into account. Testing the theory of epidemiological and health transition Vallin shows that from a comparative perspective the development of the life expectancy in Europe can be characterized by sequences of diverging and converging movements. The cross-country differentials correspond to the different onsets and developments of the epidemiological transition in the countries. Similar relationships emerge when he examines the development of intra-country differences in life expectancy by region, sex, and socio-economic categories over time. Although the epidemiological transition theory alone does not suffice to explain all such differentials, Vallin assumes that the processes of divergence and convergence brought about by different stages or distributions of economic, social, and medical transition may continue in the future. To him, reducing the causes that bring about the differences in life expectancy and mortality constitutes one of the great political challenges in Europe.

## 1.3 Migration

During the last two decades European countries have also witnessed important changes in migration trends and patterns. Migration streams have increased Sobotka (2009). Southern European countries have become a region of immigration and migration from Eastern to Western Europe has grown (Katseli et al. 2006: 13). The East–West labor migration in Europe is not a new phenomenon. East Europeans contributed to the economies of the West and North European countries as early as the late nineteenth century; some countries participated also in the post-war European labor migration. Another wave of the East–West migration took place in the late 1980s and early 1990s, both for political and economic reasons. Even though the guest worker programs underlying the labor migration of the 1960s and early 1970s were terminated, much of the ongoing labor migration within Europe is perceived as temporary (Castles 2006; Dustmann et al. 1996). The OECD estimates that between 28 and 60 % of the migrants to Western European countries return (or leave) within five years of immigration (Dumont and Spielvogel 2008: 171f). Migrants from Eastern Europe have typically short-term work contracts, and seasonal migration into agriculture, construction, and service sectors in Western Europe is common. The temporary nature of migration partly reflects the motives of migrants. Young people are willing to search income and to acquire new skills and experience abroad, which can later be used back at home. 'Western experience' and foreign language skills are seen as important dimensions of human capital valued at home. For families, working in the West offers an opportunity to accumulate savings in order to buy (or renovate) a house back at home or survive the times of economic hardship.

The question of what happens to temporary migrants after their return to their homeland has been a topic little studied in migration literature. The chapter by *David Lindstrom* helps to fill the gap by focusing on the occupational mobility of Mexican return migrants from the US, a case which shares many similarities with the intra-European migration. His analysis shows that return migrants have a higher likelihood of making occupational transitions than non-migrants in Mexico. Interestingly, however, while some migrants experience upward occupational mobility upon their return to Mexico, by contrast, another large group experiences downward occupational mobility at the time of return. The analysis of lifetime occupational mobility of migrants reveals similar patterns. While return migrants are more likely to invest in land and businesses and become self-employed than non-migrants, the US migration experience increases the likelihood of long-term downward mobility for most return migrants. The author concludes that individuals who work in the US and return to Mexico are unlikely to experience any wage return on their US work experience and may in fact be penalized for their experience. The reason may be the deterioration of location-specific human and social capital because of being away from the Mexican labor market. Employers may also discount US migration experience because they perceive return migrants as 'risky' workers who may leave again.

The study by Lindstrom focuses on the Mexico-US migration, but the similarities with ongoing East–West migration in Europe are striking. The results suggest that the experiences of the East–West migrants may not be as rosy as young people and families hope. Working in the UK, Germany or France is seen by young Eastern Europeans as an investment in human capital (including learning a major European language). However, often migrants in construction or agriculture mostly work together with their countrymen or countrywomen and have limited contact to locals. They thus have little chances to improve their language skills and learn new things. Therefore, they may return home with little 'Western' experience and without location-specific human and social capital left at home. Still, there is reason to believe that the experiences vary by migrant groups, just as the Mexican case study also suggests. While temporary agricultural and construction workers may not benefit from working in the West in the long run, other groups, e.g. medical doctors, nurses and other highly skilled professionals, may benefit much from the work experience abroad. There are indications that the circulatory migration of highly educated professionals is based on different conditions than the temporary labor migration of low skilled workers, since highly skilled professionals may have better options of leave of absence from their jobs and since they are also the target group which EU policies aim at attracting (Castles 2006). Lindstrom's contribution illustrates the need for a broader perspective on migration, which sees migration as a sequence of temporary moves over the life course. Such an approach calls for better migration data in Europe, that is, for longitudinal individual-level data with information on migration as an integral part of a person's life course. As Andersson (2004) and Kulu and Milewski (2007) have shown, only such data can also contribute to better understand the linkages between migration and other demographic events relevant for the understanding of Europe's demographic pattern.

## 1.4 Methodological Advancement

There is a consensus among demographers nowadays that a real understanding of the demographic development of Europe is not possible if one only focuses on macro-level data. Aggregate data may reflect the current demographic status and, if time series are available, they may also give some indication of change, but they cannot provide much insight into what behavior has brought about the current status or induced the change over time. The suggestion made by McDonald in this book to include duration aspects in the calculation of the TFR is only one example of the increasing acknowledgement that individual-level data are needed in order to assess demographic processes (see also: Hoem and Mureşan 2011a, b; Hoem et al. 2011a). To understand the causes and consequences of demographic change, longitudinal, individual-level data and appropriate analytical methods are regarded as necessary prerequisites (see also: Andersson and Neyer 2004; Neyer and Andersson 2008). The contributions by Mayer and Schulze and by Lindstrom illustrate the need for such data. These days, event-history models applied on the individual level are regarded as state-of-the-art methodology. In his chapter, *Niels Keiding* focuses on two issues of event-history analysis which most researchers are confronted with: the effect of non-standard sampling patterns and the concept of 'local dependence'. Both issues pose specific problems. Niels Keiding describes the development in dealing with these features and points out that there are still unresolved issues. His contribution does not only underline the progress in event-history analysis over the past decades; it also shows the need to strive for further methodological sophistication in this area.

## 1.5 Summary and Acknowledgment

The contributions in this book were collected with the purpose of addressing highly relevant issues in the study of the demographic development of Europe. The chapters demonstrate not only the diversity of demographic issues in Europe, but they first and foremost show that much needs to be done, theoretically, methodologically, and empirically, to gain a better understanding of demographic trajectories and their linkage to other societal developments. The contributions were first presented at the symposium "The Demography of Europe" organized to honor Jan M. Hoem on his retirement as Director of the Max Planck Institute for Demographic Research. The editors thank all the contributors to this book, the participants at the symposium, and the staff of the Max Planck Institute for Demographic Research for their support. The editors are very grateful to the reviewers of the different book chapters and to Karin Tesching for editorial assistance. Our special appreciation goes to Evelien Bakker, Bernadette Deelen-Mans and Sundari of the Springer Verlag for their untiring support.

Our best wishes and primary thanks go to Jan Hoem. We are deeply grateful for having had the opportunity to be part of his research staff at the Max Planck Institute for Demographic Research, for the good and productive time we had together there, and for his continual and inspiring support thereafter. Several of the research questions and issues addressed in this book were also part of the research program on contemporary European fertility and family dynamics which Jan initiated and led at the MPIDR. We are happy to see the progress in this research area and we are even more happy to see that Jan still continues to work and to contribute to the progress in research on the demography of Europe (see list below).

## References

Andersson, G. (2004). Childbearing after migration: Fertility patterns of foreign-born women in Sweden. *International Migration Review, 38*(2), 364–392.

Andersson, G., Kreyenfeld, M., & Mika, T. (2009a). Welfare state context, female earnings and childbearing in Denmark and Germany. *Stockholm Research Reports in Demography, 2009*, 13.

Andersson, G., Rønsen, M., Knudsen, L., Lappegård, T., Neyer, G., Skrede, K., et al. (2009b). Cohort fertility patterns in the Nordic countries. *Demographic Research, 20*(14), 313–352.

Andersson, G., & Neyer, G. (2004). Contemporary research on European fertility: Perspectives and developments. Edited Special Collection 3 of *Demographic Research*.

Billari, F., Goisis, A., Liefbroer, A., Settersten, R. A., Aassve, A., Hagestad, G., et al. (2011). Social age deadlines for the childbearing of women and men. *Human Reproduction, 26*(3), 616–622.

Bernardi, L., Klärner, A., & von der Lippe, H. (2008). Job insecurity and the timing of parenthood: A comparison between Eastern and Western Germany. *European Journal of Population, 24*(3), 287–313.

Castles, S. (2006). Guestworkers in Europe. A resurrection? *International Migration Review, 40*(4), 741–766.

Dumont, J.-C., & Spielvogel, G. (2008). Return migration: A new perspective. *OECD International Migration Outlook*, Part III. SOPEMI.

Dustmann, C., Bentolila, S., & Faini, R. (1996). Return migration: The European experience. *Economic Policy, 11*(22), 213–250.

European commission (2009). *Aging report: Economic and budgetary projections for the EU-27 member states (2008–2060). European economy 2/2009*. Luxembourg: Office for official publications of the European communities.

European commission (2011). *Demography report 2010: Older, more numerous and diverse Europeans*. Luxembourg: Publications office of the European Union.

Frejka, T., & Sobotka, T. (2008). Overview Chapter 1: Fertility in Europe: Diverse, delayed and below replacement. In Frejka, T., Sobotka, T., Hoem, J.M., & Toulemon, L. (Eds.), *Childbearing trends and policies in Europe. An overview. Demographic Research 19*(3): 15–46.

Goldstein, J., Sobotka, T., & Jasilioniene, A. (2009). The end of lowest-low fertility? *Population and Development Review, 35*(4), 663–699.

Hoem, J. M., Jalovaara, M., & Mureşan, C. (2011a). Recent patterns in group-specific total fertility in Finland. *Stockholm Research Reports in Demography, 2011*, 20.

Hoem, J. M., & Mureşan, C. (2011a). The total marital fertility rate and its extensions. *European Journal of Population, 27*(3), 295–312.

Hoem, J. M., & Mureşan, C. (2011). An extension of the conventional TFR. *European Journal of Population, 27*(4), 389–402.

Kantorová, V. (2004). Education and entry into motherhood: the Czech republic during state socialism and the transition period (1970–1997). *Demographic Research Special Collection, 3*(10), 246–273.

Katseli, L., Lucas, R.E.B., & Xenogiani, T. (2006). Effects of migration on sending countries. What do we know? *OECD Development Centre Working Paper* No. 250.

Keim, S., Klärner, A., & Bernardi, L. (2009). Qualifying social influence on fertility intentions: Composition, structure, and meaning of fertility-relevant social networks in Western Germany. *Current Sociology, 57*(6), 888–907.

Kreyenfeld, M. (2004). Fertility decisions in the FRG and GDR: An analysis with data from the German fertility and family survey. *Demographic Research Special Collection, 3*(11), 275–318.

Kreyenfeld, M. (2010). Uncertainties in female employment careers and the postponement of parenthood in Germany. *European Sociological Review, 26*(3), 351–366.

Kulu, H., & Milewski, N. (2007). Family change and migration in the life course: An introduction. *Demographic Research, 17*(19), 567–590.

Liefbroer, A., & Billari, F. (2010). Bringing norms back in: A theoretical and empirical discussion of their importance for understanding demographic behaviour. *Population Space and Place, 16*, 287–305.

Neyer, G., & Andersson, G. (2008). Consequence of family policies on childbearing behavior: Effects or artifacts? *Population and Development Review, 34*(4), 699–724.

Neyer, G., & Hoem, J. M. (2008). Education and permanent childlessness: Austria vs. Sweden; a research note. In Surkyn, J., van Bavel, J., & Deboosere, P. (Eds.), *Demographic challenges for the 21st century: A state of the art in demography* (pp. 91–112). Brussels: VUBPRESS Brussels University Press. Also available as MPIDR Working Paper, WP 2008–007.

Oeppen, J., & Vaupel, J. W. (2002). Broken limits to life expectancy. *Science, 296*, 1029–1031.

Sobotka, T. (2009). Migration continent Europe. *Vienna Yearbook of Population Research, 7*, 217–233.

Van Bavel, J. (2010). Choice of study discipline and the postponement of motherhood in Europe: The impact of expected earnings, gender composition, and family attitudes. *Demography, 47*(2), 439–458.

## Post-Retirement Publications of Jan Hoem

Frejka, T., Sobotka, T., Hoem, J. M., & Toulemon, L. (2008). Childbearing trends and policies in Europe. Edited Special Collection 7 of *Demographic Research*, 3 volumes. Including Preface by Hoem, J. M. (Volume 1, 1–4) and Summary and General Conclusions by the four editors (Volume 1, 5–14).

Gabrielli, G., & Hoem, J. M. (2010). Italy's non-negligible cohabitational unions. *European Journal of Population, 26*(1), 33–46.

Hoem, J. M. (2008a). The impact of public policies on European fertility. In Frejka, T., Sobotka, T., Hoem, J. M., & Toulemon, L. (Eds.), Childbearing trends and policies in Europe. Special Collection 7 of *Demographic Research,* Volume 1: 249–260.

Hoem, J. M. (2008b). The reporting of statistical significance in scientific journals: A reflexion. *Demographic Research, 18*(15), 437–442.

Hoem, J. M. (2010). 'Demography', 'The life table', and 'Statistical significance': Three contributions to Lovric, M. (Ed.), *International encyclopedia of statistical science.* Springer Verlag.

Hoem, J. M., Gabrielli, G., Jasilioniene, A., Kostova, D., & Matysiak, A. (2010). Levels of recent union formation: Six European countries compared. *Demographic Research, 22*(9), 199–210.

Hoem, J. M., Jalovaara, M., & Mureşan, C. (2011). Recent patterns in group-specific total fertility in Finland. *Stockholm Research Reports in Demography, 2011*, 20.

Hoem, J. M., & Kostova, D. (2009). Early traces of the second demographic transition in Bulgaria: A joint analysis of marital and non-marital union formation. *Population Studies, 62*(3), 1–13.

Hoem, J. M., Kostova, D., Jasilioniene, A., & Mureşan, C. (2009a). Traces of the second demographic transition in four selected countries in central and Eastern Europe: Union formation as a demographic manifestation. *European Journal of Population, 25*(3), 239–255.

Hoem, J. M., Kostova, D., Jasilioniene, A., & Mureşan, C. (2009b). The structure of recent first-union formation in Romania. *Romanian Journal of Population Studies, 3*(1), 33–44.

Hoem, J. M., & Mureşan, C. (2011a). The total marital fertility rate and its extensions. *European Journal of Population, 27*(3), 295–312.

Hoem, J. M., & Mureşan, C. (2011b). An extension of the conventional TFR. *European Journal of Population, 27*(4), 389–402.

Hoem, J. M., & Mureşan, C. (2011c). The role of consensual unions in Romanian total fertility. *Stockholm Research Reports in Demography, 2011*, 16.

Hoem, J. M., & Nedoluzhko, L. (2008). Marriage formation as an intermediary process between migration and childbearing. *Demographic Research, 18*(21), 611–628.

Mureşan, C., & Hoem, J. M. (2010). The negative educational gradients in Romanian fertility. *Demographic Research, 22*(4), 95–114.

Neyer, G., & Hoem, J. M. (2008). Education and permanent childlessness: Austria vs. Sweden; a research note. In Surkyn, J., van Bavel, J., & Deboosere, P. (Eds.), *Demographic challenges for the 21st century: A state of the art in demography* (pp. 91–112). Brussels: VUBPRESS Brussels University Press.

# Chapter 2
# Challenges for European Family and Fertility Research

**Peter McDonald**

## 2.1 Introduction

In the first decade of the twenty-first century, many countries in Europe are facing the issue that their fertility rate has remained very low for a long period of time. Very low fertility sustained for a reasonable time period, in most cases, will severely distort the age structure of the population. It will mean in the short term that education and other child-related services will have to be cut back considerably. More importantly, in the medium term, it will mean that the size of the labor force will begin to fall sharply just as many populations are reaching the peak of their ageing process (when the baby-boom generation reaches their 80s). This will put health and social security systems under severe pressure. Already, the viability of many European social insurance systems is questioned. This situation has given rise to a demand for research that will assist governments to understand the causes of low fertility and to define policy approaches to low fertility. This chapter addresses the challenges that are inherent in this research agenda.

In the field of demography, the basic tenets of classical demographic transition theory (Thompson 1929; Notestein 1945, 1953) hold considerable sway. One of its tenets is that fertility will fall from high, natural levels to an unchanging, controlled level that will see all societies having a fertility rate of two children per woman, the level that replaces the population. There is a certain planning logic to this as well. If we want populations to stop rising in number but we do not want to see them fall, then replacement level fertility is the scientific solution. The stationary population that eventuates from replacement level fertility also provides

P. McDonald (✉)
Australian Demographic and Social Research Institute, The Australian National University, Canberra, Australia
e-mail: peter.mcdonald@anu.edu.au

G. Neyer et al. (eds.), *The Demography of Europe*,
DOI: 10.1007/978-90-481-8978-6_2,

a relatively favorable age structure for planning purposes. Thus, replacement fertility is the logical long-term outcome for fertility.

With this wisdom, it was not surprising that, when it first appeared, demographers regarded below replacement fertility as a temporary aberration that would go away of its own accord. It was not an area of policy interest. A demographic explanation that is consistent with the notion that below replacement fertility is a temporary phenomenon was found for very low fertility, the 'tempo effect' (Bongaarts and Feeney 1998; Kohler and Ortega 2002; Zeng and Land 2002; Sobotka et al. 2005; Rodríguez 2006). By this explanation, very low period rates of fertility are caused by an increasing delay of births, particularly the delay of the first birth, but the completed fertility of cohorts of women remains the same, at replacement level. This complacent approach has been called into question by the emergence in some countries of completed cohort fertility levels that are well below replacement level, and still trending downwards (Frejka 2008a). The realization that below replacement fertility may become socially entrenched presents challenges for European family and fertility research. This contribution addresses some of these challenges.

## 2.2 Estimation of Future Fertility Rates

The pressure is upon European demographers to produce more reliable estimates of future fertility for planning purposes. In response to this pressure, European organizations conventionally publish 'tempo-adjusted' corrections of the conventional Total Fertility Rate and claim these as the emergent levels of fertility in European countries (Vienna Institute of Demography 2008). The tempo-adjusted Total Fertility Rate is based on the notion that the conventional period Total Fertility Rate is distorted by the increasing delay of births that has extended over a very long period of time. And, there is no question that the timing of births is a very important consideration in the projection of future fertility. However, the calculation of the tempo-adjusted Total Fertility Rate is based upon a simplistic, mathematical approach. It relies on the totally unrealistic assumption that there is no relationship between tempo and quantum, that is, that the timing of births has no influence on completed fertility. Furthermore, the adjustment parameters used in the calculation are themselves affected by tempo. For example, more recent birth cohorts may be shifting their childbearing to younger ages while, at the same point in cross-sectional time, older cohorts are having their (delayed) births at older ages. In this situation, cross-sectional measures of timing like the mean age at birth or the mean age at first birth may continue to rise for a while because of the late childbearing of cohorts aged 30 and over. In other words, the adjustment parameters are relatively insensitive to change.

The unreliability of the tempo-adjusted Total Fertility Rate as a measure of emergent fertility is evidenced by the fact that there are now 11 countries in Europe where the actual completed fertility of the 1965 birth cohort is lower than or equal to the tempo-adjusted Total Fertility Rate (Austria, Belgium, Bulgaria, Denmark, Georgia, France, Germany, Luxembourg, Netherlands, Switzerland and the United Kingdom) (Vienna Institute of Demography 2008). For Germany, the completed cohort fertility rate for the 1965 birth cohort has fallen to 1.55 births per woman. Subsequent cohorts in Germany have lower levels of achieved fertility at each age suggesting even lower levels of completed fertility in the future (Dorbritz 2008).

Simplistic measures have no place in the estimation of future fertility. There is no alternative to detailed, country-specific estimations. Age-parity cohort completion methods offer some scope in this regard. They are certainly an advance upon the tempo-adjustment approach. However, with Kippen, I have argued that the interval since the previous birth is also an important predictor of the timing and level for births beyond the first birth (McDonald and Kippen 2007, 2008a; Kippen and McDonald 2009). For Australia, using a cohort completion approach that incorporated three parameters, age, parity and interval since the previous birth, simultaneously applied, we successfully projected the turning point in Australian period fertility after 2001 (McDonald and Kippen 2008b). For Australia, we have observed a high degree of stability across decades of birth probabilities by age, parity (parity >0), and interval since last birth. We concluded that this stability needs to be incorporated in projections of future period fertility. Also important was the incorporation into the calculations of the existing population structure by age, parity and interval since the last birth. This structure builds in the impacts of past changes in the timing of births. The challenge for European fertility research is to obtain these very detailed measures of fertility for each country.

First births have been delayed to increasingly older ages in western, northern and southern European countries over the past two or more decades. It is likely that this trend has run its course such that further increases in cohort age at first birth are unlikely. This provides an ideal opportunity to use detailed information on fertility to examine how the tempo and quantum of fertility changed during this transition to a much older age at first birth (McDonald and Kippen 2009). Are there similarities across countries? What are the differences in the experiences of those countries where fertility has fallen to very low levels (under 1.5 births per woman) compared with those where fertility has been sustained at moderate levels (1.7–2.0 births per woman)?

Where fertility has fallen to low levels, the key to its projection lies in accurate estimation of the timing and level of first births. While demographic time series approaches are worthy of consideration, the use of characteristics and behavioral determinants is desirable. Thus, in projecting future births, the most important challenge for European researchers is to devise methods for the prediction of age-specific probabilities of first birth. Longitudinal panel surveys such as the Generations and Gender Surveys provide opportunities for this to be done. However, societal-level influences on the timing and level of first birth probabilities are also likely to be important. This is discussed in the next section.

Ultimately, these detailed approaches to the projection of births can only be reliable over a period of not more than about 10 years. However, for planning purposes (maternity services, early childhood education, primary school education), more reliable projection of births over a decade is highly beneficial.

## 2.3 Fertility as a Macro-Social Issue

I have argued (McDonald 2006a) that all advanced countries have been subjected to similar forces of social change over the past 30 years, particularly social liberalization and economic restructuring, but some countries have come through this period with moderately low levels of fertility (1.7–2.0 births per woman) while others have ended with very low levels of fertility (under 1.5 births per woman). This suggests that there are explanations of fertility that are institutional or structural, rather than individual alone. In the cited article, I have argued that the explanation lies in cross-national variations in the family–state dynamic, but other potential explanations are possible. Whatever the explanation, low fertility should be considered to be a macro-sociological issue as well as an individual issue. Individuals make their decisions about fertility within a given institutional context that has bearing upon their decisions. Furthermore, institutional contexts affecting fertility are changing across time.

In a very useful theoretical paper, Hans-Peter Blossfeld has stated:

> Any theoretically powerful sociological analysis of a macro-social problem must pay attention to both structural and micro-level issues (Blossfeld 1996, 181).

Very low fertility is a macro-social issue. In simple terms, Blossfeld argues that it is theoretically impossible to explain a major macro-trend such as the movement to very low fertility by only considering changes in the circumstances or characteristics of individuals. He is critical of rational choice models (commonly used in economics) in which the institutional context is taken for granted (exogenous) and is assumed to be unchanging. Clearly, a higher degree of complexity is involved. This more complex approach requires an assessment of the social institutions that impinge upon the issue under consideration (in this case, very low fertility). How do social institutions constrain or enhance an individual's decision making and to what extent are the relevant social institutions changing? This is another challenge for European fertility research.

Social institutions often change slowly and this is a reason that they may be ignored in rational choice models. An individual making a decision about a birth does so in the existing social context and takes only the existing social context or perceived short-term changes in that context into account. This tends to support rational choice theory, but if the aim is to explain the society's level of fertility rather than the individual's level of fertility, then the changing social context is clearly important. Part of the problem is that analysis is often based upon cross-sectional survey databases. In such databases, it is intrinsically impossible to ask people how they may have behaved in the social context of 10 years ago. Thus, to measure the impact of the changing social context upon decision making, longitudinal panel data are required. Also, different long-run institutional changes may 'fold over' each other as time progresses. It may be the cumulative impact of a number of institutional changes that produces the effect upon fertility decision making. Where this occurs, the impact of any particular institutional change may be very difficult to demonstrate in any empirical way even if we have longitudinal panel data over a long period of time.

We are much more willing to draw conclusions about the relevance of the social-institutional context when that context changes dramatically and swiftly. For example, fertility rates fell sharply during the depression years of the 1930s. For Australia, Ruzicka and Caldwell (1977, 120) concluded that marital fertility for women in their early 20s 'responded quite promptly to the down turn of the economy in the early 1930s'. Marriage rates also fell during the 1930s depression but commentators in the 1930s were not suggesting that the problem of the low birth rate could be solved by 'getting people to marry' or by changing their values about children. Everyone knew that the low birth rates, the low marriage rates and people's current attitudes were influenced by the one large cause, economic uncertainty induced by the economic depression, an institutional or structural change.

A more recent example of importance being given to the social-institutional context is the sharp fall in fertility in the Central and Eastern European countries after 1990. In a major book that analyzes fertility changes in European countries, Frejka unequivocally attributes the fall in fertility to the replacement of the state socialist regimes by market economies and by fledgling democratic institutions of governance:

> The former created the relatively favorable conditions for childbearing of job security, low-cost housing, free education, free health care, and a variety of entitlements associated with child birth and childrearing, as well as shortages of career opportunities, leisure activities, and consumer goods. These were replaced by the more restraining conditions for childbearing of job insecurity, an increasing pressure to acquire more education, expensive housing, lesser and declining birth and childrearing entitlements, as well as the availability of a variety of career opportunities, leisure activities and consumer goods. The populace of the CEE countries had grown accustomed to the socialist paternalistic welfare state circumstances over several decades, and, all of a sudden, young people were confronted with the need to deal with a whole new Western type of family formation and childbearing environment. Market economy principles and Western democratic institutions provided the framework for family formation and childbearing, and they also provided the basis for the adoption and diffusion of western type norms, values and attitudes regarding demographic behavior (Frejka 2008b, 166–167).

As an example of behavioral change, in the Czech Republic, Kantorová (2004) found that the shift to a market economy led to greater differentiation in labor market opportunities by education which in turn led to greater education differentiation in the timing of entry into motherhood (Kantorová 2004). It is also notable in the above quotation that Frejka attributes not only changes in individual behavior to the social-institutional changes but also changes in the values of young people in the Central and Eastern European countries. The state socialist regime provided a strong constraint to the spread of Western values. The sweeping away of the regime led to the rapid acceptance of Western value orientations. I shall return to the role of values in fertility decision making below.

Another example of an institutional impact on fertility that is widely acknowledged in the literature is the change in Swedish policy in the 1980s that promoted a speedier second birth (Hoem 1990, 1993; Andersson 1999, 2002, 2004; Andersson et al. 2006). The increase in second birth rates in Sweden following this change was nothing short of spectacular. The age standardized second birth rate in Sweden was 60 % higher in 1990 than it had been in 1977 and about 50 % higher than the 1990 age standardized second birth rates in Norway and Denmark, comparable countries that had not introduced the 'speed premium' (Andersson 2004). Of this Swedish experience, Neyer and Andersson have noted:

> The effect is of general interest since it constitutes one of the clearest demonstrated cases of a true causal effect of a social policy change on fertility behavior in a modern society (Neyer and Andersson 2007, 19).

The above examples of the impacts of institutional change upon fertility are undisputed because of their spectacular nature: sudden and dramatic institutional change followed by a very clear and definite fertility response. However, when institutional change is less spectacular and more diffuse, its impacts are more difficult to observe and this provides scope for the social-institutional context to be passed over in research. Multi-country longitudinal panel surveys provide some scope for the analysis of institutional effects and, in the European context, more quantitative studies of this type are desirable. At the same time, there is scope for more socio-political studies of the changing institutional context in individual countries and how this institutional context may have influenced the timing or level of fertility. Indeed, such studies are probably required to develop theory that may drive the quantitative studies.

## 2.4 Values Associated with Fertility Decision Making

At the individual level, there are good theoretical reasons explaining the delay in first births that has been observed over the past 30 years. Delay allows young people, especially women, the opportunity to accumulate human capital prior to the disruption that occurs with the first birth. In an increasingly competitive labor market, this strategy is highly rational and more so for the majority that are risk averse. However, eventually the delay of the first birth reaches a level that introduces new risks, in particular, the risk that the desired number of births will not eventuate or the risk that a childless relationship will break up. Many European countries are reaching or have reached this level. Some prominent women have written influential books that declared their regret at not being able to have the number of children that they wanted (Crittenden 2002; Hewlett 2002; Haussegger 2005). These books received widespread media coverage in the countries in which they were published. This all gives rise to the possibility that social attitudes in support of latter commencement of childbearing may change and that age at first birth may begin to fall on average. Questions for research in this regard are: Is this a possible direction? Is there any sign of it at present? What are/would be the determinants of a fall in age at first birth at the societal level? Lutz and Skirbekk (2005) have argued that earlier commencement of childbearing should be a policy approach of governments in Europe. Can policy make a difference?

More generally, there is a dearth of studies of child-bearing decision making and of the perceived benefits of children. Research issues here could include:

- How can the strength of the desire to have children be measured?
- Which features of children give strength to these desires?
- What differential values do children bring at the different birth orders?
- How persistent are value orientations in relation to family and fertility as circumstances change?
- How important to childbearing is the quality of the couple relationship and the views of the partner in fertility decision making?
- How do partners talk about having children?
- What are the perceived risks of having children?
- What are the perceived impacts of children upon the couple relationship?
- To what extent do these values differ across European countries?
- Do changing values provide any direct evidence of the institutionalization of very low fertility (the low fertility trap hypothesis) (Lutz et al. 2006)?
- What influences values related to having children? Is it public discussion, policy, a sense of economic insecurity, gender inequity, materialism, etc.? Can observed national differences be explained in broader theoretical terms?

Answers to these questions are likely to entail both qualitative and quantitative research approaches.

## 2.5 Theory Development

A greater degree of sophistication in theory development is required to address social change relating to family and fertility. To this time, in demography, the prominent theoretical approach has been Second Demographic Transition theory (Lesthaege and van de Kaa 1986; van de Kaa 1987). This is a unilinear and universal theory that argues that the features of the Second Demographic Transition such as cohabitation, childbearing outside marriage, divorce and low fertility will increase inexorably in any cultural setting as the values of western individualism inevitably spread into that culture. It is a theory of changing values, not of changing institutions. While not underestimating the power of western ideas about personal freedom, I have argued for a more nuanced approach to social change which recognizes a varying degree of modification of cultures in the face of western influence but adaptation that does not imply universal outcomes (McDonald 1993, 1994; Utomo and McDonald 2009). In this theoretical framework, values and their sources remain important but the institutional setting is not assumed to be easily swept aside.

For example, Japan is increasingly put forward as a case that verifies the Second Demographic Transition theory. Marriage rates and fertility have fallen and the divorce rate is rising. Both marriages and births have been delayed substantially and women's labor force participation has risen. However, it is not unusual in other societies that marriages and births have been delayed substantially and that many people do not marry, but fertility is much higher in some of these societies than it is in Japan. The Netherlands has later marriage than Japan and just as high a percentage are not in lifetime partnerships but the Netherlands has higher fertility. In France, women work much longer hours than Japanese women do but their fertility is higher. The United Kingdom has a higher divorce rate than Japan but it also has higher fertility than Japan. In general, in advanced economies, countries with high proportions of births outside of marriage have higher fertility rates than those that do not. It is well known that the countries that are further along the path of the so-called Second Demographic Transition (e.g. the Nordic countries) have higher fertility than numerous countries that are not so far along the path. It is very evident that very low fertility is not explained by Second Demographic Transition theory.

As I have argued elsewhere (McDonald 2006a), theory relating to family and fertility change needs to incorporate institutional factors especially social transformations that influence perceptions of gender equity (McDonald 2000a, b) and economic trends and orientations that affect individual perceptions of risk. The more favorable economic conditions that prevailed before the recent economic downturn may well have reversed the movement to later childbearing and lowered divorce rates. I am presently working on Australian evidence that suggests that this is the case. And it seems that this trend may have been driven by better material outcomes rather than by post materialist or postmodernist values (see van de Kaa's thought-provoking article 2001) but, at this point this is little more than speculation. Suffice to say that there is much scope in European countries for the development of theory through research in this area. Furthermore, the recent severe economic downturn provides an excellent opportunity to measure the effects of fluctuations in economic fortunes upon family and fertility behavior.

## 2.6 Approaches to Fertility and Family Policy

If, as I argue, very low fertility has institutional explanations, then it is subject to policy intervention through modification of or accommodation to the institutional causes of very low fertility. This implies research on the development of appropriate policies. Most European research to this point has consisted of evaluation of existing policies rather than the development of alternative policies. Also, most research to this point has focused on the evaluation of a specific policy (cash payments, child care, parental leave, etc.) rather than upon the impact of the full range of policies. Often, this research aims to determine what is the one, most effective policy. Is it cash payments, is it child care, is it parental leave? Most of this research is inconclusive with evidence of no or only small effects of any one policy. In reality, however, it is the coherence and consistency of the full range of family policies that is important in determining the policy impact upon fertility. An individual policy may not appear to have much effect but it may be a necessary component in a total package that provides young people with the confidence to form families. Its effect may be more evident in its absence than in its presence as seems to have been the case in the Eastern and Central European countries after 1990 (Frejka 2008b). Comprehensive studies have been conducted in Europe (d'Addio and d'Ercole 2005; Neyer 2003) but more studies of this type are required across time as policies and institutional environments change.

A more comprehensive approach to family policy research requires an audit of a nation's family policies with an assessment of each policy against a set of desirable criteria (McDonald 2002, 2003, 2006b). Once this is done, it is possible to consider the form (in theory) of a comprehensive reform package that would be feasible in the country under consideration. Implementation of a reform package requires an understanding of the politics of family policies to support fertility. What policy approaches are capable of receiving broad political support? Pronatalist policies are easily portrayed as policies of the extreme right parties in politics, and, as such, may well be taken up by rightist parties particularly those opposed to immigration. Once pronatalist policy becomes embedded in this framework, it will fail because of its divisiveness and its minority appeal. How can pronatalist policy be pursued while avoiding ownership of its agenda by the extreme right? How can a truly national approach to policy be engendered? These are important issues for research.

The fertility levels of immigrants can arise as a topic of interest in policy debate. Given the rising significance of immigrants in many European countries, more study of the demographic behavior of immigrants is desirable. Does the demographic behavior of immigrants converge to that of the non-immigrant population? Is this still the case in the second generation? What is the incidence of intermarriage? What is the demographic impact of differential immigrant behavior? These questions have been addressed recently by Sobotka (2008) but the topic will remain important in European population debate.

Another significant area for policy research is the importance of perceptions and symbols.

> Investigations into the effects of family policies on fertility need to consider both the normative or symbolic connotations of family policies and their correspondence with societal development (Neyer and Andersson 2008, 704).

There is some evidence from qualitative interviews in Australia that government policies directed to the needs of working families with children have had a symbolic value that may be more significant than their dollar value (McDonald et al. 2008). Research of this type will probably be qualitative in nature because of the difficulty of formulating adequate quantitative questions that address this issue.

## 2.7 Conclusion

To support the important research agenda that is described in this chapter, there is an evident need for better data, more timely data and better analysis of that data. First, the endeavor and resources that are applied to the production of timely economic indicators need also to be applied to indicators of fertility. Analysis of fertility trends needs to be detailed and current. This requires the collection of data on births by age, parity and duration since the previous birth and the estimation of the population according to the same characteristics. It also requires the development of measures that provide a more reliable picture of the directions in which fertility is likely to head in the future. This implies a better understanding of trends in the tempo and quantum of fertility than is the case at present.

Second, in examining the causes of low fertility, there is no substitute for cross-national longitudinal panel surveys that are able to incorporate the three main dimensions of fertility decision making: the micro-level costs and constraints, the institutional setting and the ideological influences (attitudes and values). The important impact of the institutional (and policy) setting can only be observed across time or across geographic units, especially across countries. Such databases also allow the investigation of heterogeneity in fertility decision making in each country. National policies to support families with children can miss the mark badly if there is a poor understanding of the heterogeneity of values, attitudes, costs and constraints in fertility decision making across the country.

Third, the chapter has put an emphasis on obtaining a better understanding of why people choose to have children and on the values associated with fertility decision making. It provides a list of research questions on this issue that can be used to guide new research in this area.

Fourth, all this research adds to the development of theory. The discipline of demography has been criticized in the past with some justification for being theory-free or numbers-driven. While this is gratifyingly much less the case with the younger generation of European demographers, theory should always be in the forefront of research on fertility decision making and the discipline of demography needs to devote itself more to the theoretical considerations in its field.

The final and most difficult area for research on fertility is the area of policy research. While desirable, it is difficult for ethical, pragmatic and political reasons to apply experimental designs to policy research in this area. Policy takes time to be effective. Policy is political, policy is ideological and, to an uncomfortably large extent, policy is about perceptions rather than realities. At base, countries should conduct audits of the full range of policies that impinge upon fertility decision making. These audits need to incorporate assessments of each policy dimension against appropriate policy criteria (see McDonald 2006b). However, this chapter also asserts that policies that support families with children are important in their total effect rather than in their individual effectiveness. This means that attention needs to be given to the coherence of the range of policies in the context of heterogeneity within the country. Finally, to the extent that policy impact is more about perceptions than realities, qualitative research is required to investigate how different people see or value the existing policy regime. Monitoring of public discussion of fertility and family policy also can play a part in gaining an understanding of public perceptions.

## References

Andersson, G. (1999). Childbearing trends in Sweden 1961–1997. *European Journal of Population, 15*(1), 1–24.

Andersson, G. (2002). Fertility developments in Norway and Sweden since the early 1960s. *Demographic Research, 6*(4), 67–86.

Andersson, G. (2004). Childbearing developments in Denmark, Norway, and Sweden from the 1970s to the 1990s: A comparison. *Demographic Research, 3*(7), 155–176. (Special Collection).

Andersson, G., Hoem, J., & Duvander, A.-Z. (2006). Social differentials in speed-premium effects in childbearing in Sweden. *Demographic Research, 14*(4), 51–70.

Blossfeld, H.-P. (1996). Macro-sociology, rational choice theory, and time: A theoretical perspective on the empirical analysis of social processes. *European Sociological Review, 12*(2), 181–206.

Bongaarts, J., & Feeney, G. (1998). On the quantum and tempo of fertility. *Population and Development Review, 24*(2), 271–291.

Crittenden, A. (2002). *The price of motherhood.* New York: Metropolitan Books.

d'Addio, A., & d'Ercole, M. (2005). *Trends and determinants of fertility rates in OECD countries: The role of policies.* Paris: OECD.

Dorbritz, J. (2008). Germany: Family diversity with low actual and desired fertility. *Demographic Research, 7*(17), 139–170. (Special Collection).

Frejka, T. (2008a). Parity distribution and completed family size in Europe: Incipient decline of the two-child model. *Demographic Research, 7*(4), 139–170. (Special Collection).

Frejka, T. (2008b). Determinants of family formation and childbearing during the societal transition in Central and Eastern Europe. *Demographic Research, 7*(7), 139–170. (Special Collection).

Haussegger, V. (2005). *Wonder woman: The myth of having it all.* Sydney: Allen and Unwin.

Hewlett, S. (2002). *Baby hunger: The new battle for motherhood.* London: Atlantic Books.

Hoem, J. (1990). Social policy and recent fertility change in Sweden. *Population and Development Review, 16*(4), 735–748.

Hoem, J. (1993). Public policy as the fuel of fertility: Effects of a policy reform on the pace of childbearing in Sweden in the 1980s. *Acta Sociologica, 36*(1), 19–31.

Kantorová, V. (2004). Education and entry into motherhood: The Czech Republic during state socialism and the transition period (1970–1997). *Demographic Research, 3*(10), 245–274. (Special Collection).

Kippen, R., & McDonald, P. (2009). Using census data to create a detailed database for fertility analysis. Paper to the Annual Meeting of the Population Association of America, Detroit, April 30–May 2.

Kohler, H.-P., & Ortega, J. A. (2002). Tempo-adjusted period parity progression measures, fertility postponement and completed cohort fertility. *Demographic Research, 6*(6), 91–144.

Lesthaeghe, R., & van de Kaa, D. (1986). Twee demografische transities? In D. van de Kaa & R. Lesthaeghe (Eds.), *Bevolking: Groei en Krimp* (pp. 9–24). Deventer: Van Loghum Slaterus.

Lutz, W., & Skirbekk, V. (2005). Policies addressing the tempo effect in low-fertility countries. *Population and Development Review, 31*(4), 699–720.

Lutz, W., Skirbekk, V., & Testa, M. R. (2006). The low-fertility trap hypothesis: Forces that may lead to further postponement and fewer births in Europe. *Vienna Yearbook of Population Research, 2006*, 167–192.
McDonald, P. (1993). Convergence or compromise in historical family change? In E. Berquo & P. Xenos (Eds.), *Family systems and cultural change* (pp. 15–30). Oxford: Clarendon Press.
McDonald, P. (1994). Families in developing countries: Idealised morality and theories of family change. In L. Cho & M. Yada (Eds.), *Tradition and change in the Asian family* (pp. 19–28). Honolulu: East-West Center.
McDonald, P. (2000a). Gender equity, social institutions and the future of fertility. *Journal of Population Research, 17*(1), 1–16.
McDonald, P. (2000b). Gender equity in theories of fertility transition. *Population and Development Review, 26*(3), 427–439.
McDonald, P. (2002). Sustaining fertility through public policy: The range of options. *Population, 57*(3), 417–446.
McDonald, P. (2003). Reforming family support policy in Australia. *People and Place, 11*(2), 1–15.
McDonald, P. (2006a). Low fertility and the state: The efficacy of policy. *Population and Development Review, 32*(3), 485–510.
McDonald, P. (2006b). An assessment of policies that support having children from the perspectives of equity, efficiency and efficacy. *Vienna Demographic Yearbook, 2006*, 213–234.
McDonald, P., & Kippen, R. (2007). The intrinsic total fertility rate: A new approach to the measurement of fertility. Paper to the Annual Meeting of the Population Association of America, New York.
McDonald, P., & Kippen, R. (2008a). Delay of first births in Australia, 1981–2001: Its nature and its impact upon the timing and incidence of second births. Paper to the Annual Meeting of the Population Association of America, New Orleans, April 17–19.
McDonald, P., & Kippen, R. (2008b). The projection of births: a new approach. Paper to the Conference of the Australian Population Association, Alice Springs, June 30–July 3.
McDonald, P., & Kippen, R. (2009). Measuring the quantum of fertility during a long-term shift from early to late childbearing: Australia 1946–2005. Paper to the International Population Conference, Marrakech, Sept 27–Oct 2.
McDonald, P., Evans, A., Gray, E., & Kippen, R. (2008). Policy and fertility: The case of Australia, 1996–2006. Paper to the IUSSP Seminar on Fertility and Public Policies in Low Fertility Countries, Barcelona, Spain, July 7–8.
Neyer, G. (2003). Family policies and low fertility in Western Europe. *Supplement to Journal of Population and Social Security, 1*, 46–93.
Neyer, G., & Andersson, G. (2007). *Consequences of family policies on childbearing behavior: effects or artifacts? MPIDR WP 2007–021*. Rostock: Max Planck Institute for Demographic Research.
Neyer, G., & Andersson, G. (2008). Consequences of family policies on childbearing behavior: Effects or artifacts? *Population and Development Review, 34*(4), 699–724.
Notestein, F. (1945). Population—the long view. In T. Schultz (Ed.), *Food for the World* (pp. 36–57). Chicago: University of Chicago Press.
Notestein, F. (1953). Economic problems of population change. In: *Proceedings of the Eighth International Conference of Agricultural Economists* (pp. 13–31). London: Oxford University Press.
Rodríguez, G. (2006). Demographic translation and tempo effects: An accelerated failure time perspective. *Demographic Research, 14*(6), 85–110.
Ruzicka, L., & Caldwell, J. (1977). *The end of demographic transition in Australia*. Canberra: The Australian National University.
Sobotka, T., Lutz, W., & Philipov, D. (2005). 'Missing births': Decomposing the declining number of births in Europe into tempo, quantum, and age structure effects. European Demographic Research Paper, Vienna Institute of Demography.

Sobotka, T. (2008). The rising importance of migrants for childbearing in Europe. *Demographic Research, 7*(9), 139–170. (Special Collection).

Thompson, W. (1929). Population. *American Journal of Sociology, 34*(6), 959–975.

Utomo, I., & McDonald, P. (2009). Adolescent reproductive health in Indonesia: Contested values and policy inaction. *Studies in Family Planning, 40*(2), 133–146.

van de Kaa, D. (1987). Europe's second demographic transition. *Population Bulletin, 42*(1), 1–59.

van de Kaa, D. (2001). Postmodern fertility preferences: From changing value orientation to new behavior. *Supplement to Population and Development Review, 27*, 290–311.

Vienna Institute of Demography. (2008). *European Demographic Data Sheet 2008*. Vienna: Vienna Institute of Demography.

Zeng, Y., & Land, K. C. (2002). Adjusting period tempo changes with an extension of Ryder's basic translation equation. *Demography, 39*(2), 269–285.

# Chapter 3
# Welfare States, Family Policies, and Fertility in Europe

**Gerda Neyer**

## 3.1 Fertility Developments in Europe

Over the last 40 years of the twentieth century, period fertility developments in Europe were characterized by two specific trends, namely the decline of fertility to below the replacement level (i.e. to a Total Fertility Rate below 2.1) and the movement from intraregional diversity to intraregional homogeneity (Neyer 2003). The decline of the TFR affected all of Europe. In many countries, the TFR dropped to very low levels (below 1.5) or even to lowest-low levels (below 1.3). Some countries experienced the decline in the late 1970s and early 1980s, others encountered a drop only in the 1990s. At the end of the twentieth century, all European countries had a TFR below the replacement level, and in roughly two-thirds of the countries the TFR was below 1.5. Overall, about three-quarters of the European population lived in societies with such a low fertility (Frejka and Sobotka 2008, 17).

During the first decade of the twenty-first century, the TFR rose again in almost all major European regions. By 2008, no European country had a lowest-low fertility level and only eight countries had a TFR below 1.4. Another eight countries even reached highest-low fertility levels (TFR 1.85 or above; Eurostat 2010, Fig. 3.1). The increase proved not to be the same across Europe. Contrary to developments in most of Western Europe, the TFRs of the German-speaking countries (Germany and Austria) as well as of neighboring Switzerland failed to increase from their very low levels, while in some high-fertility countries (France, Sweden, the UK) the increase was larger in absolute terms than what many other countries experienced (Goldstein et al. 2009, 671). In Central and Eastern Europe, the onset and the range of the increase varied greatly (Goldstein et al. 2009, 669).

G. Neyer (✉)
Stockholm University Demography Unit, Department of Sociology, Stockholm University, Stockholm, Sweden
e-mail: gerda.neyer@sociology.su.se

G. Neyer et al. (eds.), *The Demography of Europe*,
DOI: 10.1007/978-90-481-8978-6_3,

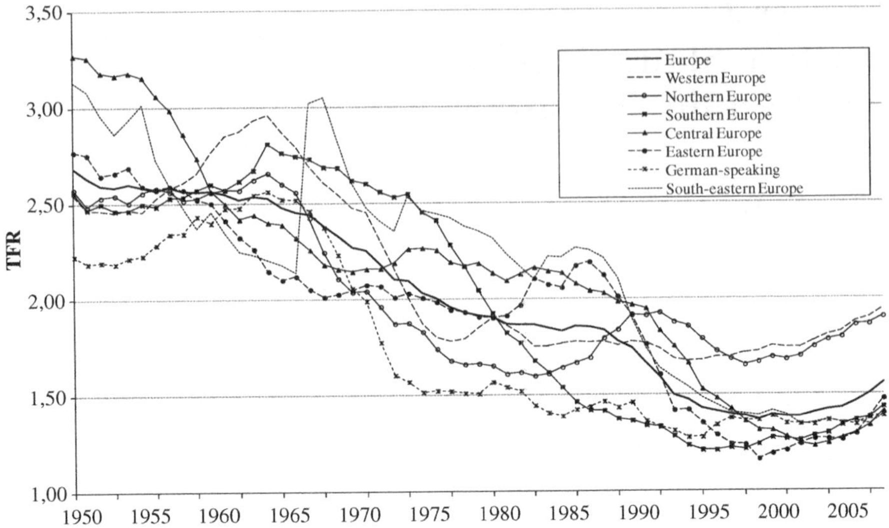


**Fig. 3.1** Total Fertility Rate (TFR) in major regions of Europe, 1950–2008 *Source*. Frejka and Sobotka 2008, 21; Sobotka (unpublished data up to 2008). *Notes* Data are weighted by the population size of given countries and regions. Countries are grouped into regions as follows: Western Europe: Belgium, France, Ireland, Luxembourg, the Netherlands, and the United Kingdom; German-speaking countries: Austria, Germany, and Switzerland; Northern Europe: Denmark, Finland, Iceland, Norway, and Sweden; Southern Europe: Greece, Italy, Portugal, and Spain; Central Europe: Croatia, Czech Republic, Estonia, Hungary, Latvia, Lithuania, Poland, Slovakia, Slovenia; South-eastern Europe: Bulgaria, Macedonia, Romania, Serbia and Montenegro; Eastern Europe: Belarus, Moldova, Russia, and Ukraine; Europe: EU-27, ex. Cyprus and Malta

With the subsequent economic crisis, many affected countries lost their gains in the TFR (Sobotka et al. 2011).

The general fertility decline of the last four decades of the twentieth century was accompanied by a process of regional convergence. In the 1960s, Europe's fertility pattern was rather dispersed and showed marked differences in fertility levels even between countries in the same region. By the beginning of the twenty-first century, fertility levels had moved towards greater homogeneity within regions. At the same time fertility differences between regions had not diminished; they had remained stable or had even increased slightly (Neyer 2003). Due to the uneven increase in the TFR at the beginning of the twenty-first century the regional disparity between the high-fertility regions (the Nordic countries, France, the Netherlands, Ireland and the UK) and the low-fertility regions (the German-speaking countries, Southern, Central and Eastern Europe) has thus persisted or has even become more pronounced (Frejka and Sobotka 2008; Goldstein et al. 2009).

Low fertility and distinct regional differences are not a new phenomenon in Europe. During the first decades of the twentieth century, fertility in Western and

Eastern Europe also declined to below replacement levels. By the mid-1920s, more than half of the European population lived in countries with sub-replacement fertility (Frejka and Ross 2001, 214f.), and by the mid-1930s only a minority of European countries had fertility rates above replacement (van Bavel 2010, 2). The fertility increase to and beyond replacement level that followed in the late 1940s and during the "baby boom" of the 1950s and 1960s marks a temporary phase in the otherwise low-fertility history of twentieth-century Europe.

Despite the long history of low-fertility patterns in Europe, comparisons and subsequent conclusions about sub-replacement fertility have usually been drawn with the high-fertility decades of the mid-twentieth century as a reference point. Just like in the 1920s and 1930s, the currently low fertility in Europe has generated several doomsday scenarios that predict the demise of the continent. Europe has been pictured as graying and even as vanishing demographically. Some authors have seen it as sliding into a global disadvantage due to its lack of young people and have portrayed it as falling behind in technological advancement, and as losing out in productivity, economic performance, and political power. Many authors fear that low fertility and the shortage of young people will have adverse effects on Europe's labor markets and on European social-security and welfare-state systems. Immigration, proposed by some as a remedy against the consequences of low fertility, is rejected by others who warn that further immigration will jeopardize Europe's social and cultural cohesion, its ethnic homogeneity, and the survival of "Europeanness". (For a summary of these arguments, see Neyer 2011; for the perception of low fertility in the 1920s and 1930s, see van Bavel 2010.)

Such concerns about the consequences of low fertility have spurred much public and political interest in policies that could raise fertility levels. The increase in the TFR during the first decade of the twenty-first century has not silenced the voices that call for such policies. Quite on the contrary, the number of countries which state that they endorse policies to elevate fertility has continued to rise since the turn of the century. In 2001 about a third of the European countries reported that they had implemented policies to raise fertility. By 2009, half of the countries claimed to have such policies in place (United Nations 2002, 2010). The European Union has also come to view low fertility as a challenge. It has started a number of initiatives to monitor the demographic trends of Europe and to guide countries in taking policy measures which are assumed to increase fertility and to promote "demographic renewal" (Policies 2006).

To an observer, the national and supra-national initiatives to tackle low fertility levels raise the question whether policies do in fact influence fertility developments and bring about or at least contribute to higher fertility. Many studies which compare the fertility-related policies and the TFR of selected countries attribute differences in fertility levels to differences in policies supporting families (Sleebos 2003; Castles 2003; Gauthier 2007). Unfortunately, the results stated depend on the methods used (Hoem 2008), on the selection of countries, and on the time frame in which the TFR and the policies are compared. Some European countries with comparatively high levels of fertility have very meager policies (e.g., Ireland or the UK); other countries with extensive family support programs have experienced continued or temporary low fertility levels (e.g., the German-speaking countries; Sweden in the first half of the 1990s). This challenges the unconditional assumption that fertility-related policies boost fertility levels.[1] Even the recent increase in the TFR in many low-fertility countries in Europe cannot unanimously be attributed to fertility-related policy initiatives enacted to increase fertility. According to Goldstein et al. (2009), in some countries the rise in the TFR was probably related to the implementation of such policies, in others there is no discernible link between family policy initiatives and fertility developments, and in yet another group of countries the fertility turn-around happened despite the absence of any major fertility-related policy intervention. In general, researchers conclude that the effects of policies on general fertility levels are weak, inconclusive, or contradictory (Neyer 2003; Sleebos 2003; Gauthier 2007; Thévenon and Gauthier 2011; Spiess 2012; see also the country chapters in Frejka et al. 2008).

This has led some researchers to take a broader perspective on policies. They propose that one should pursue an institutional approach (McNicoll 1994) and consider the configurations and interactions of policies, explore the impact of policy packages, and take the welfare-state system into account (McNicoll 2001; Neyer 2003, 2005; Neyer and Andersson 2008; McDonald in this book). Others argue that such a broadening of the angle complicates the assessment of policy effects on fertility because it is hardly possible to observe and measure the impact of policy conglomerations or of the welfare state on human lives (Mayer and Schoepflin 1989; Rindfuss and Brauner-Otto 2008). I maintain in this chapter that the broader policy- or welfare-state approach on the one hand and the single-policy approach on the other hand complement each other. Concentrating on the welfare state, I outline the potential of an integrative approach for fertility research and present empirical examples to demonstrate its analytical range. The main conclusion of this exercise is that both approaches are necessary in order to explain fertility developments in Europe from a policy perspective. Together, they provide a more integrative framework which allows us to link policies and fertility both theoretically and empirically.

## 3.2 Welfare State and Fertility

Several researchers have pointed out the importance of the state for fertility developments (McNicoll 1994, 2001; Mayer and Schoepflin 1989; Mayer 2009; McDonald 2006). They stress that the capacity of the state to shape fertility derives from its legal, administrative, and bureaucratic power to regulate vital relationships within society (McNicoll 1994) and to structure the individual's life course (Mayer 2009). European countries are unique in that the welfare state has become

[1] The same applies to policies which aim at lowering fertility (Bledsoe et al. 1998).

an essential institution—some maintain even *the* essential institution—which shapes human behavior and social and economic relationships. Although many proclaim that the "golden age of the welfare state" (i.e., from the 1950s to the mid-1970s) has passed and that economic and demographic challenges will lead to the dismantling of the welfare state, current research does not corroborate this claim (Pierson 1996, 2001; Boeri et al. 2001). The welfare-state reforms in European countries during the most recent three decades have rather been what Pierson (2001) calls recalibrations of the welfare state. Many of the adjustments have involved the modification of existing and the introduction of new welfare policies (Pierson 2001, 16). They signal a shift in the focus of the welfare state towards greater responsiveness to "new social risks" (Esping-Andersen 1999; Bonoli 2005; Palier 2010). These "new social risks" are largely related to the transformation of gender relationships and to the decline of the "male-breadwinner family" (Esping-Andersen 1999, Esping-Andersen et al. 2002; Bonoli 2005; Taylor-Gooby 2004). They involve primarily woman-centered life-course risks which have been covered insufficiently by the "old" welfare state with its focus on the income retention of the male breadwinner, viz, the reconciliation of work and family life, care needs, and lone parenthood (Bonoli 2005, 431). These "new social risks" are closely related to childbearing. The same consideration applies to other "new social risks" to which welfare states have become more responsive, namely, low education, low income, long-term unemployment, and insufficient social-security coverage (Bonoli 2005; Esping-Andersen 1999). This implies that the welfare state has become potentially more important for childbearing decisions and fertility developments. It makes the question of how and to what extent we can assess the impact of the welfare state on fertility even more pressing.

### 3.2.1 Quantifying the Welfare State

Researchers have pursued several strategies to measure the importance of the welfare state for human life. One approach is to classify welfare states according to how extensive or generous they are. Most often generosity or extensiveness are measured in monetary or other quantifiable terms. Examples are the share of the general domestic product attributed to social policies, expenditure on family-relevant social policies (e.g., health), the length of maternity leave and of parental leave after childbirth, the wage compensation granted to mothers (and fathers) during such leaves, the benefits granted to single mothers or to each child in a family, public subsidies for childcare, housing, education, and so forth. The attraction for demographers of these and similar indicators is that they are often available for a large number of countries and that many of these indicators have been standardized for comparative purposes. Examples of such databases are: the OECD Social Expenditure Database, the OECD Family Database, the European System of Integrated Social Protection Statistics (ESSPROS), the Comparative Family Policy Database, and the GGP Contextual Database. During the last two decades, more fine-grained indicators have become available, as have previously missing ones that are associated with the welfare coverage of "new social risks", of gender, and of family issues. This means that demographers have an increasing number of and more refined indicators at hand. Such indicators are well suited to rank welfare states or to determine clusters of welfare states along demography-relevant aspects (see, for example, Thévenon 2008). A weakness of such indicators is that many of them represent only the quantitative dimensions of a welfare state and its policies. Merely quantifying welfare-state measures have proved to be of limited potential in capturing essential characteristics of the welfare state and in explaining current or long-term cross-national differences in fertility outcomes. Even if measures of specific policies are selected, relying on the purely quantitative aspects of such policies may not be sufficient to show possible linkages between policies and fertility. The reason for this is that extensiveness or generosity does not necessarily indicate how supportive a policy or a welfare state is. For example, social expenditure may be high because of high unemployment, which in turn leads to greater expenditure on unemployment benefits, even if these are low. Likewise, a long parental leave appears quantitatively as generous, while in fact long leaves are known to inhibit mothers' return to the labor market and women's employment careers (Rønsen and Sundström 2002). A merely quantitative approach does not uncover such features, because it neither discloses the principles on which a welfare state is based nor its aims. Both of these latter dimensions are essential, however, in that they shape the content and the direction of a welfare state and its social policies (for an example of an approach which recognizes these dimensions, see Kalwij 2010). They are therefore crucial elements needed to capture the essence of the welfare state and to assess its potential impact on fertility.

### *3.2.2 The Welfare State from the Perspective of its Principles and Aims*

Several researchers have developed welfare-state concepts that recognize explicitly the principles and the aims of a welfare state (Wilensky 1975; Flora and Heidenheimer 1982; Esping-Andersen 1990). Such notions have become dominant in welfare-state research. This approach differs significantly from the purely quantifying approach. The latter regards a welfare state primarily as the "sum of its policies" (measured, as we mentioned above, in terms of extensiveness or generosity). An approach which considers the principles and aims of a welfare state focuses on the structuring functions of the welfare state and on the ideological premises on which its social policies are based (Wilensky 1975). It regards the welfare state as an institution which provides social order for human behavior through its social policies. As an institution, it creates a "stable equilibrium" (McNicoll 2009, 790) for individual action and for society, and it gives guidance and meaning to behavior.

## Welfare Regimes: The Principles of Equality and of Security

The principles and aims acknowledged as essential for welfare states are commonly related to equality and to security. Equality is usually conceived as equality across class lines. Some distinguish between equality of opportunities and equality of result (Flora and Heidenheimer 1982). Security relates primarily to protection against economic and demographic risks (unemployment, old age) through social-security systems (Flora and Heidenheimer 1982). Based on the way in which these principles and aims are transposed into welfare-state policies, distinct welfare regimes can be identified. In his seminal and widely applied typology of welfare regimes, Esping-Andersen (1990) classifies Western welfare states according to three dimensions which govern core social policies, related to unemployment, health/invalidity, and old-age pensions:

1. "De-commodification", which is "the degree to which individuals, or families, can uphold a socially acceptable standard of living independently of market participation" (Esping-Andersen 1990, 37). In essence, this refers to the social security granted to a person in case of unemployment, sickness, old age, or any other life-course issue which makes her/him unable to participate in the labor market (e.g., childbearing, childrearing).
2. "Social citizenship", which is whether access to welfare-state benefits are granted on the basis of individual rights (social citizenship), on the basis of needs (Esping-Andersen 1990, 21–23), or on the basis of specific family constellations. Benefits based on needs often rest on the principle of subsidiarity, in that the family is regarded as the primary source of welfare and the state only as a back-up. Benefits are only granted if individual and family resources are exhausted or if the family is unable to provide welfare. Entitlements based on specific family forms (marriage) acknowledge the principle of mutual support in family relationships and grant access to social security to the partner/children of the main provider. In both cases, an individual's welfare is made dependent on family relationships and/or family resources (Fraser and Gordon 1994).
3. "Stratification", which is whether welfare-state policies reinforce, sustain, or reduce social cleavages and status differences between various social and economic groups. This includes whether access to and levels of benefits are granted on equal terms to everyone or whether they vary by occupational status (blue-collar/white-collar/public workers) or by family status (married/non-married).

These dimensions generate three specific types of Western welfare regimes: (1) the *universalistic* or *social-democratic* welfare regime of the Nordic countries, characterized by a high degree of de-commodification, social-citizenship based rights to welfare benefits, and policies which are directed towards social equality; (2) the *conservative-corporatist* welfare regime of continental Western Europe, with stratifying and subsidiary welfare systems based on employment- and occupation-related social insurance schemes; and (3) the *liberal* welfare regime of the English-speaking countries with its reliance on the market and on private provision of social security and with minimal (means-tested) public support.

There have been several suggestions to modify Esping-Andersen's typology (see, Arts and Gelissen 2002 for an overview). Most of them are based on claims that specific countries deviate from the ideal types of welfare regimes. Following Ferrera (1996), many researchers, including demographers, acknowledge Southern European countries as a separate welfare regime (the "Latin rim" or "Mediterranean" model), because of the highly fragmented system of social protection, the close reliance on the family as provider of welfare, the distinct clientalism in these countries, and the pronounced labor-market insider–outsider divide.[2] Post-communist countries in Eastern and Central Europe have been added as an additional, although diverse and volatile, welfare regime, which combines Bismarckian (conservative) and liberal (market-based) welfare aspects (Cerami and Vanhuysse 2009). All of these modifications of the three worlds of welfare maintain the primary focus on the core social-security systems in Western Europe (unemployment, health/invalidity, old-age pension) and on their logics regarding social rights, protection against labor-market risks, and class structures.

## Welfare Regimes, Gender, and the Family

Feminist and gender researchers have broadened this framework further by emphasizing the notions of gender ingrained in welfare policies. They emphasize how views of gender shape welfare-state policies and how welfare states in turn shape gender relationships in the family, in the economic sphere, and in society. They thus include the family in a more subtle way in their classification and analysis of welfare states. They de-compose "the family" and distinguish between partnership and parenthood as two different forms of relationships which are shaped by welfare-state policies. They introduce care, that is, care giving and care receiving, as essential dimensions of welfare policies (Knijn and Kremer 1997). They bring in family policies as part of social policies. Finally, they modify the welfare-state principles by turning the attention to risks, rights, and stratifications related to or resulting from gender (being female or male), sexual orientation (heterosexual, same-sex, transsexual), partnership forms (cohabitation, marriage, divorce, widowhood), or parenthood status (single mothers; other forms of parental status). (For an overview over these aspects, see Neyer 2003).

Taken together these features expand the range and content of social policies substantially. Just like the labor market or class structure, the family, gender, and sexual relationships are regarded as targets of the welfare state. Linking gender, the labor market, the family, and the welfare state, feminist and gender welfare-

[2] The allocation of the Netherlands and of Switzerland to the conservative welfare regime has also been contested; both are seen as hybrid cases (Arts and Gelissen 2002).

state researchers have highlighted three additional fertility-relevant welfare-state principles:

1. "Commodification", which is the way in which welfare states support the labor-force participation of women and mothers, and maintain or alter the "male-breadwinner" family (Orloff 1993; Lewis 1992; Gornick et al. 1997). In a broader sense, commodification includes maintaining and supporting the employability of individuals, especially of those who have difficulties entering and remaining in the labor market (e.g., mothers, youth, immigrants, the less educated, the long-term unemployed). It also extends to the way in which policies promote or hamper gender equality in the labor market, that is, in types of employment, at the workplace, in income, in career options and so forth, as an essential aspect of commodification over the life-course.
2. "Familialism"/"de-familialisation", which depicts the way in which a welfare state deals with "care". "Familialism" refers to the extent to which welfare states regard the family and thus directly or indirectly women as the primary providers of care. It describes the degree to which welfare states support familial care, encourage mothers' retreat from the labor market after childbearing, and strengthen their caring functions (Leitner 2003). This includes the preferential treatment of marriage (and of a married female carer) over other family forms. "De-familialisation" refers to the aim of a welfare state to reduce family care responsibilities and (female) dependencies by providing institutional care (Anttonen and Sipilä 1996; Esping-Andersen 1999; Saraceno and Keck 2010; Saraceno 2011).
3. "De-gendering of employment and care", which is the extent to which welfare states seek to alter gender relationships by supporting the employment of mothers and by promoting the engagement of fathers in childcare and family work (Sainsbury 1999; Korpi 2000). This refers primarily to policy aims to change the male-breadwinner/female-carer family model towards the more gender-equal dual-earner/dual-carer family form (Sainsbury 1999).

If the dimensions of commodification, familialism, and de-gendering of employment and care are added in the classification of welfare states, the welfare-regime pattern becomes more diverse. The continental European welfare states do not constitute a common regime any longer. With their high degree of de-familialisation, France and Belgium are closer to the Nordic welfare states. Among the other continental welfare states, there exist different patterns of support for women's employment, of familialism and de-familialisation: Southern Europe is less de-familialising than other conservative Western European countries; Eastern Europe, once a region in which states ideologically supported female employment and encouraged childbearing through various degrees of familialism, has undergone a process of re-familialisation and moved towards different, albeit frequently changing patterns of familialism (Lewis 1992; Anttonen and Sipilä 1996; Gornick et al. 1997; Sainsbury 1999; Leitner 2003; Neyer 2005; Szelewa and Polakowski 2008; Szikra and Tomka 2009).

## Family-Policy Change and Regime Continuity

The family-policy based pattern of welfare regimes has become more diverse during the past decade. The reforms of the European welfare states in order to tackle new social risks—in particular policy initiatives related to (female) labor-force participation, parental leave, and childcare—have not been uniform across countries nor have they all gone in the same direction. For example, Finland, Norway, and recently also Sweden introduced some "re-familialisation" policies by giving parents a choice between using public childcare or taking care of children at home for a longer period of time. These countries seem to partly move away from their gender-equality principles, so that the Nordic welfare regime appears less homogenous with respect to de-familialisation. Germany, by contrast, replaced its conservative parental-leave system by regulations modeled after the Nordic countries. Judged on the basis of its current parental-leave system, it emerges as a de-familialising country, deviating from the rest of the group of countries with conservative parental-leave patterns. Austria took a position in-between by diversifying its parental-leave regulations so that parents can now choose among five different modes of leave. Some authors find that such policy changes, which diverge from the general welfare orientation of a country, have made it more difficult to allocate countries to specific welfare regimes and to locate stable patterns. Some countries even tend to shift between different regimes, depending on which policy is included in the analysis. (See, e.g., the classification of care regimes in Saraceno and Keck 2010 and of family policies in Thévenon 2011). This may also make it more difficult to assess linkages between welfare-state configurations and fertility.

However, such fuzziness and instability arise primarily if the classification is based on only one single policy at a time. Radical, paradigmatic changes in policy directions are rare (Hall 1993). Even if such changes do occur, they usually apply to only one policy or to a very limited set of policy features. The change usually does not affect the direction of the welfare state (although it can mark the beginning of a gradual transformation). If we consider the main principles and orientations of welfare- and family policies as a whole, the regime patterns have remained remarkably persistent over time (Castles and Obinger 2008). Despite the changes mentioned above, Germany and Austria, for example, continue to be conservative welfare states, and the Nordic countries still constitute a distinct, highly gender-equality oriented welfare regime. Fundamental changes of the welfare state as such have happened only in post-communist countries. They were not brought about by a replacement of one set of welfare policies through another, but by the breakdown of the political and economic system, the concurrent disruption of the organizational and administrative basis of much of the (non-family) welfare production, and the rise of capitalism. These developments "dislodged" previous welfare policies and welfare functioning (Mahoney and Thelen 2010). However, some basic principles, such as familialism, have continued to prevail and have governed post-communist welfare adjustments in a way that some authors see more path-dependency and path-continuation than path-departure and path-cessation, despite the transformation of the welfare systems (Cerami and Vanhuysee 2009).[3]

## Welfare Regimes and Fertility Research

For several reasons, the lines of welfare-state research described above provide a useful approach for policy-oriented fertility research. *First*, this approach links welfare-state policies more closely to human behavior than purely expenditure-based approaches do. By focusing on the notions of gender, equality, security, class, social rights, and family incorporated in welfare-state policies, this research emphasizes that welfare states are not neutral towards all groups of society and towards all forms of social behavior. Welfare states support specific types of behavior and discriminate against others. They give privilege to specific forms of life-course organizations by making work and family behavior which is compatible with welfare-state policies more meaningful, "rational" or advantageous than other forms of behavior. Welfare states thus create a realm of "agency inequality" (Korpi 2000) along gender, class, age and other social lines.

*Second*, focusing on the principles and aims governing welfare policies offers a common basis to characterize the main orientation of a welfare state, to reduce the heterogeneity of its social policies to essential features, and to cluster welfare states into broader groups of welfare regimes. This facilitates comparative research on the linkage between social policies and fertility. It is usually impossible to include all social policies in comparative research. Even if we only consider a selection of policies, the breadth of variation in policy manifestations across countries may result in findings that cannot be interpreted in a meaningful way. Welfare-state regimes represent the basic characteristics of the social-policy environment in which women and men make their fertility decisions. They can thus serve as an indicator of this environment in fertility analyses.

*Third*, by considering the welfare-state regime, we avoid over- or underestimating the effects of single policies. Policies may have different effects depending on whether they are in line with or go against other welfare-state policies and thus line up with or counteract the general orientation of a welfare state (Neyer and Andersson 2008). Singling out one policy as an explanatory factor for fertility change or for the persistence of fertility patterns may lead to wrong conclusions as to the magnitude or even the direction of a policy effect (Neyer and Andersson 2008). Situating the policy within its broader welfare-state context, that is, in relation to the main principles and aims of the welfare state, mitigates this problem.

*Fourth*, the gender and feminist welfare-state approach connects well with two major demographic theories of fertility development and fertility differentials in

[3] Authors who stress path-dependency see the post-communist welfare institutions as partly reaching back to pre-communist times (Cerami and Vanhuysse 2009).

advanced societies, namely McDonald's theory of gender equity and Goldscheider's theory of a two-step gender revolution. Both theories link low fertility in highly-advanced capitalist countries to the discrepancy between high female labor-force participation and low gender equity in family care work, that is, to the increasing "commodification" of women on the one hand and the persistence of a high degree of familialism and a low degree of de-gendering of care on the other hand (McDonald 2000; Goldscheider 2000). Both theories hold that the decline from high to low fertility levels in Western countries has been associated with an increase in women's presence in the labor market and in other public institutions and with a concomitant persistence of women's sole responsibility for family work and care, while a subsequent increase from low to higher fertility levels necessitates a stronger involvement of men in family work and care. Welfare states play a mediating role in shaping these movements via their support or non-support of gender equality in employment and care (Neyer 2005). More generally speaking, the theories predict higher fertility if the orientation of a welfare state towards women and men complies with prevailing societal gender norms and the behavior of women and men, and the theories expect fertility to decline and remain low if there is a divergence between the gender logic of a welfare state on the one hand and the societal development regarding gender norms and gender behavior on the other hand. The same may apply to the childbearing developments among specific groups in society, since some social groups hold attitudes and are in a position to lead a life course which complies with the existing social policies while other social groups are not (see also Bourdieu 1996).

So far, the theoretical assumptions about the relationship between welfare-state logics and fertility developments have primarily focused on women; this is why the gender and feminist approach has proved to be especially useful for fertility research. However, if we want to link men's fertility behavior and male fertility developments to welfare-state configurations, the gender and feminist notions may apply only partially. For men, fertility decisions still seem to depend largely on employment and on the economic security to maintain a family rather than on care options or on the possibility to combine employment and care (Neyer et al. 2011). As a consequence, typologies of welfare states centered on employment and social security, such as those by Esping-Andersen and his successors, may be more suitable to assess the linkage between welfare states and male fertility. Similar considerations may apply if, for example, we look at connections between welfare-state systems and immigrants' fertility behavior or at the interrelationship between economic developments, welfare-state configurations and fertility behavior. Briefly, to assess the relationship between welfare-state systems and fertility, it is necessary to consider which welfare-state dimensions are relevant for the persons or groups and for the event studied.

*Fifth* and partly resulting from the arguments set out above, because of the enduring nature of welfare-state principles, the regime concept offers a useful policy framework for comparative analysis of fertility developments over time. The difficulties one encounters in a cross-sectional comparative analysis which intends to include all fertility-relevant social policies are aggravated if one attempts to incorporate the development of these policies across many countries and over time. As we have mentioned, welfare-state typologies offer a way to reduce the range and variety of social policies to a few shared characteristics and to explicitly acknowledge the social-policy system either in the analysis or in the interpretation. A welfare-state approach is particularly appropriate if policy explanations are sought for trends in aggregate (cohort or period) fertility measures.[4] These can rarely be related to the introduction, existence or elimination of a single policy, such as parental leave, family benefits, or childcare institutions. Instead, aggregate fertility measures tend to reflect the broader circumstances under which childbearing decisions are made, that is their context. Regime concepts offer a structure to this context by representing essential welfare-state features which guide behavior over the life course.

To summarize, the welfare-regime approach reaches beyond purely measurable aspects of social policies and aims to capture the essence of structuring elements of welfare systems. This has its advantages, as outlined above, but it also has some disadvantages for fertility research. Some may regard it as a drawback that some welfare-state classifications are based on researchers' judgments about policy directions. This introduces a qualitative element into the analysis. A challenge is also posed by welfare states which deviate from the ideal-type welfare regimes, that is, welfare states whose policies follow rather heterogeneous principles, such as Switzerland or the former Eastern-Bloc countries. Their allocation to a specific welfare regime or—in the case of the former communist countries—their clustering into one or two welfare regimes must be considered carefully. Some authors argue furthermore that changes in the labor market, such as the increase in new types of work which are less well covered by welfare regulations, have decreased the power of the welfare state to shape individual lives, even though welfare-state policies to cover "new social risks" have been expanded (Scharpf and Schmidt 2000). Despite these issues, the concept of welfare regimes constitutes a valuable contribution to an institutional approach in fertility research by linking individual behavior with the structured macro-level environment in which a person lives.

### *3.2.3 Welfare States and Institutional Complementarities*

The institutional approach of welfare-state research stresses that welfare states do not only affect employment and family behavior, but that they also structure the labor market and other institutions. Demographers have mostly turned their attention to the relationship between welfare-state policies and family behavior.

[4] Aggregate fertility measures have received much criticism because they do not reflect fertility-relevant behavioral aspects (e.g., Hoem 2008; Hoem and Mureşan 2011). Some authors therefore consider them inappropriate for the assessment of the relationship between policies and fertility. Nevertheless, they are still the dominant indicators of fertility development and are commonly used as reference categories in policy discussions.

Less attention has been paid to the fact that different welfare regimes are associated with differently organized labor markets, as is indicated by Esping-Andersen's concept of welfare regimes based on class- and labor relations. The organization and structure of the labor market *per se* can affect family behavior, including fertility developments (Rindfuss and Brauner-Otto 2008). For example, women working in a labor market with a high share of women and with welfare-state policies which support women's employment probably encounter different childbearing conditions than women working in a labor market with a low share of female employees and with welfare-state policies which promote exit from the labor-market when a woman enters motherhood. Beyond the mere employment structure of the labor market, scholars highlight the fact that different welfare regimes offer different schemes and levels of protection to wage earners (Esping-Andersen 1990; Hall and Soskice 2001). In conservative welfare states, the various groups of employees (such as public employees, blue-collar workers, white-collar workers, workers in new types of employment such as call centers, the new self-employed, and so forth) are usually covered by distinctly different labor-protection schemes. In social-democratic welfare regimes, the protection is more universal; in liberal welfare states labor protection is largely absent. This creates labor markets of different characters, with different conditions of entry and exit, of internal mobility, of inclusion and exclusion, and of economic and social security. In a manner similar to varieties of welfare regimes, researchers therefore speak of "varieties of capitalism" (Hall and Soskice 2001).

The approach based on varieties of capitalism distinguishes between the "coordinated market economies" of Western continental and social-democratic welfare states on the one hand and the "liberal market economies" of the liberal welfare states on the other hand (Hall and Soskice 2001).[5] In contrast to the situation in liberal market economies, labor-market entry is more institutionalized in coordinated market economies. In the latter, the labor markets show a high degree of occupational and gender segregation; they offer less opportunity for upward mobility for women, and the comparatively strong social protection of workers leads to less movement into and out of the labor market (Mandel and Semyonov 2006).

Scholars using varieties of capitalism stress that the institutional complementarities are not limited to linkages between welfare-state regimes and labor-market regimes; they also extend to other societal institutions, such as to educational systems. Estévez-Abe et al. (2001) show that different market economies have different systems of skill formation. In coordinated market economies with their

[5] There is a substantial difference between the welfare-state approach and the approach based on "varieties of capitalism". The former stresses the role of the state, the latter stresses the role of firms and employers in the development of social policies and in structuring economic relationships. My application of the varieties-of-capitalism approach selects only a specific aspect of this research line, a line which to me seems useful for fertility research.

high protection of workers against dismissal, educational systems provide high-level and often highly standardized industry-, firm-, or occupation-specific skills through vocational schools and apprenticeships. In liberal market economies with their low protection and high mobility of workers, the educational systems provide general, transportable skills (Estévez-Abe et al. 2001, Estévez-Abe 2005, 2009; Hall and Soskice 2001; Soskice 2005).

The varieties-of-capitalism approach has been criticized for lumping together the universalistic welfare states of the Nordic countries and the corporatist-conservative welfare states of continental Europe (McCall and Orloff 2005; Mandel and Shalev 2009; Rubery 2009; Thelen 2012). Recent modifications of the coordinated market economies suggest that we should distinguish between "sector-coordinated" market economies with industry-based systems of protection (corresponding largely to the corporatist-conservative welfare states) and "national-coordinated" market economies with greater egalitarian aims ("egalitarian capitalism", comprising the Scandinavian welfare states) (Thelen 2012).[6] These distinctions offer a more suitable framework to capture varieties in female labor-force participation, gender (in)equality in employment, transitions into/out of the labor market, and differences in systems of skill acquisitions.

With these extensions, the varieties-of-capitalism approach offers intriguing new aspects for demographic research. It stresses the interdependency and the complementarity between different institutions, in particular between the labor market, the educational system, and welfare-state policies. If different configurations of the welfare state are associated with different systems of skill acquisition, with different circumstances in the labor market, and with different constraints regarding mobility in and out of employment and across occupations, we cannot only focus on family policies as the essential institutional force determining childbearing behavior. We also need to consider the constraints imposed by these other complementary institutions. Combining welfare-regime approaches with a varieties-of-capitalism approach leads to a more comprehensive concept of institutional setting, namely one which puts the emphasis on the linkages between the various fertility-relevant institutions. If we use a conceptual framework for fertility research which considers institutional complementarities rather than focusing on one institution only, we are better able to assess to what extent changes in one institution (e.g., the labor market or the gender system) affect the potential impact of other institutions (e.g., the welfare state, family policies) on fertility. Such a dynamic model of institutional complementarities expands the institutional framework for explaining differentials in childbearing behavior and fertility outcome which cannot be explained sufficiently well otherwise.

[6] The terms "sector-coordinated" and "national-coordinated" economies refer to the way in which labor negotiations between employers' associations and unions (e.g., collective bargaining) are carried out. The sector-coordinated economies lead to more occupation-/sector-specific welfare and labor-market systems, national-coordinated economies to more egalitarian systems (Thelen 2012).

### 3.2.4 Welfare State, Institutional Complementarities, and Fertility: An Example

To summarize the arguments and to illustrate the usefulness of such notions for fertility research, let us consider findings of studies on ultimate childlessness by educational field and educational level among Swedish and Austrian women born in 1955–1959. (For details, see Hoem et al. 2006; Neyer and Hoem 2008). In Sweden as well as in Austria, the overall level of childless among women of this cohort is the same (15.7 %). In both countries, women educated for teaching or health professions remain childless less often than women in most other educational fields, at each level of education (Figs. 3.2 and 3.3). However, Sweden and Austria differ markedly with regard to childlessness by educational level. Austrian women with an upper-secondary or tertiary education have much higher childlessness than women with a lower education, while there exist no such differences by educational level among Swedish women[7].

The differences and similarities in levels of childlessness between and among Austrian and Swedish women cannot be explained by a single factor only, such as by differences in family policies. We need a broader and more comprehensive framework, because ultimate childlessness is measured at the end of a woman's reproductive life (at age 40+). We thus observe the final outcome of a series of decisions and behaviors that stretch over more than 25 years. We argue that the patterns are produced by the distinct institutional settings of the two countries during the period in question, namely the structure and the flexibility of the educational system, the differences in labor-market and employment conditions for different educational groups, the gendered patterns in education and in the labor market, the welfare-state orientation and the gender and family policies, as well as the way in which education, labor market, and family and gender policies are intertwined. To support the argument, we consider some of the institutional commonalities and differences, concentrating on the aspects discussed above. (For more detailed descriptions, see Hoem et al. 2006; Neyer and Hoem 2008).[8] Austria and Sweden share a sufficient number of features to make a comparison of the selected cohort of women interesting. Both countries revised their educational systems in the early 1970s to facilitate access to higher education for all social groups. Both introduced parental-leave schemes and individual taxation in the first half of the 1970s to increase female labor-force participation. Both countries are classified as "coordinated market economies" with a high degree of institutionalization of labor- and social-policy issues; both countries have educational systems with a large share of vocational training; and both countries are "strong"

---

[7] There is an exception for the most highly qualified women in Sweden, that is, women with a PhD. The Austrian data do not distinguish between the different levels of university-based tertiary education, so that no comparison is possible for this most highly educated group of women.

[8] The description focuses on the period from 1970 to 2000; this covers roughly the main reproductive years of women of the cohort 1955–59.

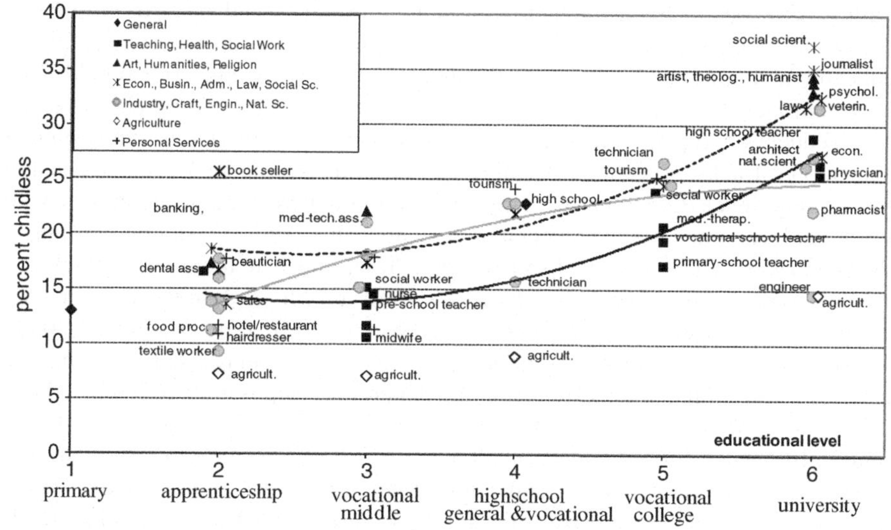


**Fig. 3.2** Percent childless by educational field and educational level. Austrian women born 1955–1959. *Source* Neyer and Hoem (2008)

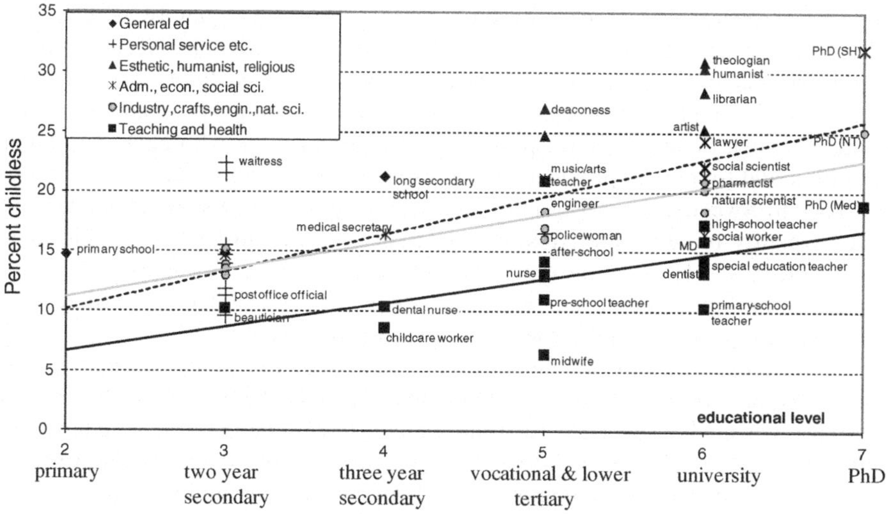


**Fig. 3.3** Percent childless by educational field and educational level. Swedish women born 1955–59. *Source* Hoem et al. (2006)

welfare states with a long-established system of family policies (Estévez-Abe et al. 2001; Culpepper 2007; Pontussen 2000; Soskice 2005). However, Austria and Sweden have taken markedly different paths regarding these institutions. Since the 1960s, Swedish policies for the labor market, social relations, education, and the family have been directed towards reducing inequality and towards promoting social and gender equality. The educational system is set up to promote higher educational attainment, to allow for flexible moves between educational fields, and to encourage recurrent and life-long education and training. Comprehensive primary education (up to age 16) avoids early separation of pupils into different educational tracks. Vocational training is offered at the (upper-)secondary and at the tertiary level. Contrary to many other coordinated market economies with vocational training systems, Sweden offers a high share of its vocational training at the tertiary level (29.9 %; Culpepper 2007, 621), and it is therefore ranked as a country with a more general educational-skill profile and high labor market mobility (Culpepper 2007, 620f; Estévez-Abe et al. 2001; Benner and Vad 2000).

By contrast, the Austrian school system is highly segregated and changes between educational tracks are difficult. Separating pupils into different educational tracks starts as early as after four years of schooling (at age 10) and continues afterwards through diversification of the school system into different types of schools or different lines of education with partially different curricula. Vocational training is offered mostly at the secondary level, either in vocational schools or as apprenticeships. (Only 3.9 % are provided at the tertiary level; Culpepper 2007, 621). Unlike in Sweden, the Austrian system of apprenticeships is segregated from the general school system. It offers much less general education and much less portable skills than the Swedish vocational system does (Culpepper 2007). There are few opportunities to interrupt and subsequently re-enter the educational system; below university level it is almost impossible. Access to post-secondary and tertiary education is restricted to those who have passed a final comprehensive high-school (upper-secondary school) exam. This leads to a class divide between those who have passed this exam and those who have not. Adult education and training is provided through specific institutions and is often costly. Contrary to Sweden, the Austrian system is not organized to support gender and social equality in education. Austria has largely adhered to gender-specific lines and programs (at least at the time when our cohorts went to school); it has not succeeded in diminishing social and class differences in educational participation and educational attainment to the extent that Sweden has.

The different educational strategies of Sweden and Austria are associated with very different patterns of educational attainment among the women of the cohorts 1955–59. Only 13 % of these Austrian women have attained tertiary education, compared to 33 % of the Swedish women. The vast majority (80 %) of the Austrian women have only completed compulsory school, vocational "middle" school, or an apprenticeship as their highest educational attainment. This compares with 53 % of the Swedish women who have equivalent degrees. Austrian women with a high-school (upper-secondary school) or higher degree constitute a clear minority (20 %), while almost half of the Swedish women hold such a degree. One can expect that these differences in educational distribution have repercussions on the labor market structure, on women's employment, and on their childbearing behavior.

The Swedish approach to the labor-market has focused on "commodification", labor-market mobility and gender equality (Benner and Vad 2000). Since the

1960s it has been directed towards increasing female labor-force participation, towards integrating every adult person into the labor market, and towards reducing gender and social inequality in pay and other work-related aspects. Welfare-state and family policies, both based on social-citizenship rights, have been geared towards this goal, as well. Since the middle of the 1970s, family policies aim to support the dual-earner/dual-carer family (Sainbury 1999), in which both women and men pursue employment and share childrearing. Parental-leave regulations have been gender neutral since then, and parental-leave benefits replace a high share of a woman's or man's previous income. Since 1995, Sweden also has reserved one month of parental leave for each parent (the "daddy month").[9] Public childcare has been extended to grant every child access to high-quality and affordable childcare. As a consequence of these policy strategies, female labor-force participation rates ranged from 60 to 83 % during 1970–2000, which is far above the corresponding rates in Austria (45–62 %).

As a conservative welfare state with a sector-/branch-coordinated welfare- and labor-policy system, Austria has put much less emphasis on integrating women into the labor market and on promoting equality in employment and care. For a long time and through most of the reproductive years of our cohort, the social-security system backed the male-breadwinner family model by granting free coverage to married (later also to non-marital registered) non-employed partners. Labor-market policies and work agreements distinguished between women and men. Direct discriminatory treatment was only gradually abolished, but indirect gender-unequal regulations have continued to exist (e.g., the seniority principle in the public sector, which has been the main employer of highly educated women). Moreover, the protection of a person's occupational and educational status in case of unemployment hampers transitions between occupations with different qualification requirements. Welfare and family policies have been directed towards facilitating mothers' work interruption after childbirth. Until 1990, parental leave, inflexible and lowly paid, was only available for mothers; the introduction of a (subsidiary) father's leave in 1990 was accompanied by a doubling of the parental-leave period to two years, which has lengthened the labor-market absence of mothers. Childcare for children below age three has hardly been available, and although the overall coverage for children at three to six years has been relatively high, there has been great regional variation regarding childcare coverage and full-time offers. In general, while Sweden has aimed at increasing female education and labor-force participation and at easing mothers' care work, Austria has maintained much of its class-oriented, gender-differentiating work/family policies, despite the fact that education has been increasing among women.

If we interpret the results of the study of Swedish and Austrian women's ultimate childlessness in light of the differences in the educational system, the

---

[9] This was extended to two months in 2002, but (a) most of the women of our cohort (1955–1959) had finished their childbearing by then, and (b) our data did not include the year 2002.

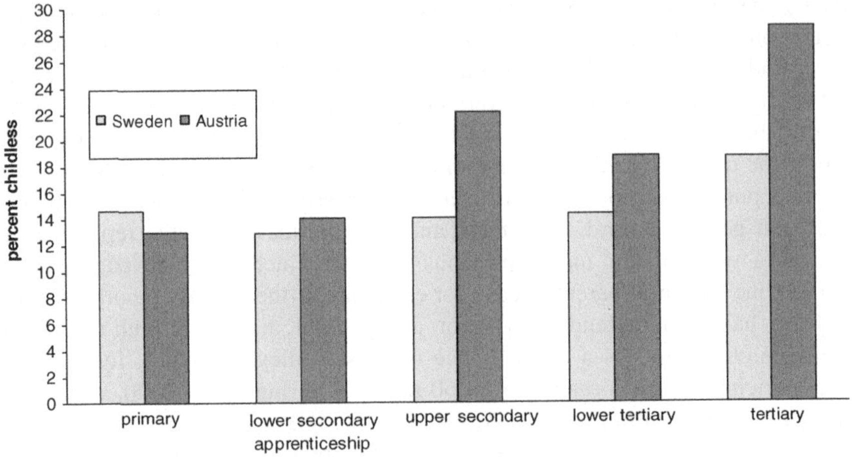


**Fig. 3.4** Childlessness by educational level at age 40+. Women born in 1955–1959: Austria and Sweden. *Source* Neyer and Hoem (2008)

labor market, the family policies, and the gender orientation of the welfare state, it is difficult to single out one factor which could explain the striking difference in the levels of childlessness between women with less education and women with higher education in the two countries (Fig. 3.4). The clear class divide in Austria indicates a discrepancy between the work aspirations of highly educated women on the one hand and the obstacles of combining work and motherhood on the other hand; the latter is brought about by the general conservative orientation of the welfare state, and the resulting conditions of education, of family-policy support, and of labor-market circumstances for women.

## 3.3 Conclusion

In this chapter, I have outlined some of the central concepts of the welfare state and have highlighted their usefulness for fertility research. The main conclusion is that studies on the impact of family policies on fertility should consider the welfare-state context in which they are embedded. Such an approach will lead to a more nuanced assessment of policy influences and provide a more rooted understanding of estimation results. In addition, it allows scholars to judge the potential fertility effects that policies might have if the social or economic environment changes or if the policies are transferred to other welfare regimes. The broader range of interpretations that the welfare-state approach offers is particularly appropriate in evaluations of the current changes of family policies in contemporary Europe and their influence on fertility.

During the past two decades family policies have become a highly dynamic policy area as many European welfare states have amended their family policies to tackle "new social risks" and to meet EU targets of raising women's labor-force participation, increasing childcare availability, facilitating work-life balance, and achieving greater gender balance in work and care. Governments also seek "best practices" in family policies, that is, options that have had the desired effect in other countries, including the effect of raising fertility. These features have greatly increased research opportunities for demographers. The changes in parental-leave and other fertility-related policies and in childcare provisions in many countries allow us to study (a) the fertility effects of similar policies in similar welfare states, (b) the effects of different policies in similar welfare states, (c) the effects of similar policies in different welfare states, and (d) the effects of different policies in different welfare states. This can be extended to explorations of the interplay between welfare regimes and economic and social developments as well as family policies, and of their influence on fertility. Such "most-similar" and "most different" research designs can produce more robust results as to the impact of policies on fertility across welfare regimes and across different economic and social environments. They allow us to better assess the influence that welfare-state and other institutional configurations as well as family policies can have on the short- and long-term development of fertility.

**Acknowledgments** I thank Tomáš Sobotka and Tomas Frejka for Fig. 3.1 and Tomáš Sobotka for the additional data for Fig. 3.1, as well as Jan Hoem and Gunnar Andersson for editorial advice. I acknowledge financial support by the Linnaeus Center on Social Policy and Family Dynamics in Europe (SPaDE), grant 349-2007-8701 of the Swedish Research Council.

## References

Anttonen, A., & Sipilä, J. (1996). European social care services: Is it possible to identify models? *Journal of European Social Policy, 6*(2), 87–100.

Arts, W., & Gelissen, J. (2002). Three worlds of welfare capitalism or more? A state-of-the-art report. *Journal of European Social Policy, 12*(2), 137–158.

Benner, M., & Vad, T.B. (2000). Sweden and Denmark: Defending the welfare state. In Scharpf, F.W. and Schmidt, V.A. (Eds.). *Welfare and work in the open economy. Volume II: Diverse responses to common challenges in twelve countries* (pp. 399–466). Oxford: Oxford University Press

Bledsoe, C., Banja, R., & Hill, A. G. (1998). Reproductive mishaps and Western contraception: An African challenge to fertility theory. *Population and Development Review, 24*(1), 15–57.

Boeri, T., Börsch-Supan, A., & Tabellini, G. (2001). Would you like to shrink the welfare state. A survey of European citizens. *Economic Policy, 16*(32), 9–50.

Bonoli, G. (2005). The politics of the new social policies: Providing coverage against new social risks in mature welfare states. *Policy and Politics, 33*(3), 431–449.

Bourdieu, P. (1996). On the family as a realized category. *Theory, Culture and Society, 13*(3), 19–26.

Castles, F. G. (2003). The world turned upside down: Below replacement fertility, changing preferences and family-friendly public policy in 21 OECD countries. *Journal of European Social Policy, 13*(3), 209–227.

Castles, F. G., & Obinger, H. (2008). Worlds, families, regimes: Country clusters in European and OECD area public policy. *West European Politics, 31*(1–2), 321–344.

Cerami, A., & Vanhuysse, P. (Eds.). (2009). *Post-communist welfare pathways. Theorizing social policy transformations in central and Eastern Europe*. New York: Palgrave MacMillan.
Culpepper, P. D. (2007). Small states and skill specificity. Austria, Switzerland, and interemployer cleavages in coordinated capitalism. *Comparative Political Studies, 40*, 611–637.
Esping-Andersen, G. (1990). *The three worlds of welfare capitalism*. Princeton: Princeton University Press.
Esping-Andersen, G. (1999). *Social foundations of postindustrial economies*. Oxford: Oxford University Press.
Esping-Andersen, G., Gallie, D., Hemerijk, A., & Myers, J. (2002). *Why we need a new welfare state*. Oxford: Oxford University Press.
Estévez-Abe, M. (2005). Gender bias in skills and social policies: The varieties of capitalism perspective on sex segregation. *Social Politics, 12*(2), 180–215.
Estévez-Abe, M. (2009). Gender, inequality and capitalism: The "varieties of capitalism" and women. *Social Politics, 16*(2), 182–191.
Estévez-Abe, M., Iversen, T., & Soskice, D. (2001). Social protection and the formation of skills. A reinterpretation of welfare state. In P. Hall & Soskice, D. (Eds.). *Varieties of capitalism. The institutional foundations of comparative advantage* (pp. 145–183). Oxford: Oxford University Press.
Ferrera, M. (1996). The 'southern model' of welfare in social Europe. *Journal of European Social Policy, 6*(1), 17–37.
Flora, P., & Heidenheimer, A. (1982). The historical core and changing boundaries of the welfare state. In P. Flora & A. Heidenheimer (Eds.), *The development of welfare states in Europe and America* (pp. 17–34). New Brunswick: Transaction Publishers.
Fraser, N., & Gordon, L. (1994). A genealogy of dependence: Tracing a keyword of the U.S. welfare state. *Signs, 9*(2), 309–336.
Frejka, T., & Ross, J. (2001). Paths to subreplacement fertility: The empirical evidence. In R.A. Bulatao & J.B. Casterline (Eds.), *The global fertility transition. Supplement to Population and Development Review* (Vol. 27) pp. 213–254.
Frejka, T. & Sobotka, T. (2008). Overview chapter 1: Fertility in Europe: Diverse, delayed and below replacement. In T. Frejka., T. Sobotka., J.M. Hoem & L. Toulemon (Eds.), *Childbearing trends and policies in Europe. Demographic Research* (Vol. 19(3), pp. 15–36).
Frejka, T., Sobotka, T., Hoem, J.M,. & Toulemon, L (Eds.). (2008). *Childbearing trends and policies in Europe*. Volume 1–3. *Demographic Research* 19.
Gauthier, A. H. (2007). The impact of family policies on fertility in industrialized countries: A review of the literature. *Population and Policy Review, 26*, 323–346.
Goldscheider, F. K. (2000). Men, children and the future of the family in the third millennium. *Futures, 32*, 525–538.
Goldstein, J., Sobotka, T., & Jasilioniene, A. (2009). The end of "lowest-low" fertility? *Population and Development Review, 35*(4), 663–699.
Gornick, J. C., Meyers, M. K., & Ross, K. E. (1997). Supporting the employment of mothers: Policy variation across fourteen welfare states. *Journal of European Social Policy, 7*(1), 45–70.
Hall, P. A. (1993). Policy paradigms, social learning, and the state: The case of economic policymaking in Britain. *Comparative Politics, 25*(3), 275–296.
Hall, P.A. & Soskice, D. (2001). An introduction to varieties of capitalism. In P.A. Hall & D. Soskice. (eds.), *Varieties of capitalism. The institutional foundations of comparative advantage* (pp. 1–70). Oxford: Oxford University Press.
Hoem, J.M. (2008). The impact of public policies on European fertility. In: T. Frejka., T. Sobotka., J.M. Hoem.,& L. Toulemon. (Eds.), *Childbearing trends and policies in Europe. Demographic Research* (Vol. 19(10), pp. 249–259).
Hoem, J. M., & Mureşan, C. (2011). An extension of the conventional TFR. *European Journal of Population, 27*(4), 389–402.

Hoem, J. M., Neyer, G., & Andersson, G. (2006). Education and childlessness. The relationship between educational field, educational level, and childlessness among Swedish women born in 1955–59. *Demographic Research, 14*(15), 331–380.

Kalwij, A. (2010). The impact of family policy expenditure on fertility in Western Europe. *Demography, 47*(2), 503–519.

Korpi, W. (2000). Faces of inequality: Gender, class, and patterns of inequalities in different types of welfare states. *Social Politics, 7*(2), 127–189.

Knijn, T., & Kremer, M. (1997). Gender and the caring dimension of welfare states: Towards inclusive citizenship. *Social Politics, 4*(3), 328–361.

Leitner, S. (2003). Varieties of familialism. The caring function of the family in comparative perspective. *European Societies, 5*(4), 353–375.

Lewis, J. (1992). Gender and the development of welfare regimes. *Journal of European Social Policy, 2*(3), 159–173.

Mahoney, J., & Thelen, K. (2010). A theory of gradual institutional change. In J. Mahoney & K. Thelen (Eds.), *Explaining institutional change: Ambiguity, agency, and power* (pp. 1–37). Cambridge: Cambridge University Press.

Mandel, H., & Semyonov, M. (2006). A welfare state paradox: State interventions and women's employment opportunities in 22 countries. *American Journal of Sociology, 111*(6), 1910–1949.

Mandel, H., & Shalev, M. (2009). Gender, class, and varieties of capitalism. *Social Politics, 16*(2), 161–181.

Mayer, K. U., & Schoepflin, U. (1989). The state and the life course. *Annual Review of Sociology, 15*, 187–209.

Mayer, K. U. (2009). New directions in life course research. *Annual Review of Sociology, 35*, 413–433.

McCall, L., & Orloff, A. S. (2005). Introduction to special issue of *Social Politics*: "Gender, class, and capitalism". *Social Politics, 12*(2), 159–169.

McDonald, P. (2000). Gender equity theory in fertility transition. *Population and Development Review, 26*(3), 427–439.

McDonald, P. (2006). Low fertility and the state: The efficacy of policy. *Population and Development Review, 32*(3), 485–510.

McNicoll, G. (1994). Institutional analysis of fertility. *Population Council Research Division Working Paper* 72.

McNicoll, G. (2001). Government and fertility in transitional and post-transitional societies. In: R.A. Bulatao & J.B. Casterline. (Eds.), *The global fertility transition. Supplement to Population and Development Review* 27, Supplement, pp. 129–159.

McNicoll, G. (2009). Legacy, policy, and circumstance in fertility transition. *Population and Development Review, 35*(4), 777–794.

Neyer, G. (2003). Family policies and low fertility in Western Europe. Supplement to Volume 1, *Journal of Population and Social Security*, 46–93.

Neyer, G. (2005). Family policies in Western Europe. Fertility policies at the intersection of gender, employment and care policies. *Österreichische Zeitschrift für Politikwissenschaft*, 91–102.

Neyer, G. (2011). Should governments in Europe be more aggressive in pushing for gender equality to raise fertility? The second "No". *Demographic Research, 24*(10), 225–250.

Neyer, G., & Andersson, G. (2008). Consequences of family policies on childbearing: Effects or artifacts? *Population and Development Review, 34*(4), 699–724.

Neyer, G., Lappegård, T., Vignoli, D. (2011). Gender equality and fertility: Which equality matters? *Stockholm Research Report on Demography* SRRD_2011:9.

Neyer, G. and Hoem, J.M. (2008). Education and permanent childlessness: Austria vs. Sweden. A research note. In J. Surkyn., Deboorsere, P. & Van Bavel, J. (Eds.), *Demographic challenges for the 21st Century. A state of the art in demography* (pp. 91–112). Brussels: VUP Press.

Orloff, A. S. (1993). Gender and the social rights of citizenship: The comparative analysis of gender relations and welfare states. *American Sociological Review, 58*(3), 303–328.

Palier, B. (Ed.). (2010). *A long goodbye to Bismarck? The politics of welfare reform in continental Europe*. Amsterdam: Amsterdam University Press.

Pierson, P. (1996). The new politics of the welfare state. *World Politics, 48*(2), 143–179.

Pierson, P. (2001). Coping with permanent austerity. Welfare state restructuring in affluent democracies. In P. Pierson (Ed.), *The new politics of the welfare state* (pp. 410–456). Oxford: Oxford University Press.

Policies to reconcile labor force participation and childbearing in the European Union (2006). *Population and Development Review 32*(2), 389–393.

Pontusson, J. (2000). Labor market institutions and wage distribution. In T. Iversen., J. Pontussen & D. Soskice. (Eds.), *Employment and central banks. macroeconomic coordination and institutional change in social market economies* (pp. 292–330). Cambridge: Cambridge University Press.

Rindfuss, R. & Brauner-Otto, S. (2008). Institutions at the transition to adulthood: Implications for fertility tempo in low fertility settings. *Vienna Yearbook of Population Research*, 2008, 57–87.

Rønsen, M., & Sundström, M. (2002). Family policy and after birth employment among new mothers: A comparison of Finland, Norway and Sweden. *European Journal of Population, 18*(2), 121–152.

Rubery, J. (2009). How gendering the varieties of capitalism requires a wider lens. *Social Politics, 16*(2), 192–203.

Sainsbury, D. (Ed.). (1999). *Gender and welfare state regimes*. Oxford: Oxford University Press.

Saraceno, C. (2011). Childcare needs and childcare policies: A multidimensional issue. *Current Sociology, 59*(1), 78–96.

Saraceno, C., & Keck, W. (2010). Can we identify intergenerational policy regimes in Europe? *European Societies, 12*(5), 675–696.

Scharpf, F. W., & Schmidt, V. A. (Eds.). (2000). *Welfare and work in the open economy. Volume I: From vulnerability to competitiveness in comparative perspective*. Oxford: Oxford University Press.

Sleebos, J. (2003). Low fertility rates in OECD countries: Facts and policy responses. *OECD Labour Market and Social Policy Occasional Papers* 15.

Sobotka, T., Skirbekk, V., & Philipov, D. (2011). Economic recession and fertility in the developed world. *Population and Development Review, 37*(2), 267–306.

Soskice, D. (2005). Varieties of capitalism and cross-national gender differences. *Social Politics, 12*(2), 170–179.

Szelewa, D. K., & Polakowski, M. P. (2008). Who cares? Changing patterns of childcare in Central and Eastern Europe. *Journal of European Social Policy, 18*(2), 115–131.

Szikra, D., & Tomka, B. (2009). Social policy in East Central Europe: Major trends in the twentieth century. In A. Cerami & P. Vanhuysse (Eds.), *Post-communist welfare pathways. Theorizing social policy transformations in Central and Eastern Europe* (pp. 17–34). New York: Palgrave MacMillan.

Spiess, C.K. (2012). Zeit, Geld, Infrastruktur und Fertilität: Befunde empirischer Mikrostudien und was wir daraus lernen können. In H. Bertram & M. Bujard (Eds.), *Zeit, Geld, Infrastruktur—zur Zukunft der Familienpolitik* (Vol. 19, pp. 321–336). *Soziale Welt, Sonderband.*

Taylor-Gooby, P. (Ed.). (2004). *New risk, new welfare. The transformation of the European welfare state*. Oxford: Oxford University Press.

Thelen, K. (2012). Varieties of capitalism: Trajectories of liberalization and the new politics of social solidarity. *Annual Review of Political Science 15*, 137–159. First published online as a review in advance 2011.

Thévenon, O. (2008). Family policies in Europe: Available databases and initial comparisons. *Vienna Yearbook of Population Research, 2008*, 165–177.

Thévenon, O. (2011). Family policies in OECD countries: A comparative analysis. *Population and Development Review, 37*(1), 57–87.

Thévenon, O., & Gauthier, A. H. (2011). Family policies in developed countries: A 'fertility booster' with side effects. *Community, Work and Family, 14*(2), 197–216.

Wilensky, H. (1975). *The welfare state and equality. Structural and ideological roots of public expenditure*. Berkeley: University of California Press.

United Nations. (2002). *National Population Policies 2001*. New York: United Nations Publications.

United Nations. (2010). *World Population Policies 2009*. New York: United Nations Publications.

Van Bavel, J. (2010). Subreplacement fertility in the West before the baby boom: Past and current perspectives. *Population Studies, 64*(1), 1–18.

# Chapter 4
# Delaying Parenthood in East and West Germany: A Mixed-Methods Study of the Onset of Childbirth and the Vocabulary of Motives of Women of the Birth Cohort of 1971

**Karl Ulrich Mayer and Eva Schulze**

## 4.1 Introduction

In this chapter, our goal is to help explain the delay in family formation among women who were born in 1971, and who live in East and West Germany. By family formation, we mean the process which leads to first births via non-marital unions and/or marriages. Delays in first births are seen as one cause of very low fertility and the increase in permanently childless couples. In particular, we want to examine whether East and West German women of the first post-transition generation still differ in regard to the age of onset of childbirth, and how they subjectively account for delays.

Our data come from the German Life History Study (Mayer 2008), in particular from two nationally representative retrospective surveys, including a panel follow-up, and from narrative interviews with respondents from the same samples. The primary material which we will use here consists of three parts: (i) the quantitative data from the retrospective surveys and the panel, (ii) biographical interviews, and (iii) the corresponding cases of the same persons reconstructed from the quantitative surveys. In addition, as a description we will show results from Kaplan-Meier estimations and Cox regressions for the whole of the East and West German 1971 cohort samples, and compare them to cohorts born between 1930 and 1960 or 1964.

Median ages at first marriage and first birth have been rising in West Germany for more than three decades, and have been increasing in East Germany since 1991. According to period data, the overall median age at marital first birth in the

---

K. U. Mayer (✉)
Leibniz Gemeinschaft, Berlin, Germany
e-mail: mayer@leibniz-gemeinschaft.de

E. Schulze
Berlin Institute for Social Research (BIS), Berlin, Germany
e-mail: e.schulze@bis-berlin.de

G. Neyer et al. (eds.), *The Demography of Europe*,
DOI: 10.1007/978-90-481-8978-6_4,

whole of Germany reached its lowest point in 1970, at age 24, and has since risen beyond age 29. By 2006, period fertility had converged in East and West Germany (Dorbritz 2008; Tivig and Hetze 2007: 18). This unprecedented delay in the onset of family formation (and thus in the conclusion of the transition to adulthood) has been attributed to educational expansion and the implicit lengthening of time in education and training (Blossfeld and Huinink 1991; Huinink and Mayer 1995; Kreyenfeld 2006), uncertainties in the labor market (Bernardi et al. 2008; Kurz 2008), a decline in collectivist and material values, and a rise in individualistic values (Lesthaege and Surkyn 1988). In addition, some observers have pointed to a shift towards more hedonistic and consumption-oriented values and lifestyles, and have noted that increasing education and labor market integration among women have led to conflicts related to women's greater career investments (Brewster and Rindfuss 2000), the lack of adequate child care, and the rise in non-marital unions (Meyer and Schulze 1992). For East Germany, three additional causes must be considered: (a) the influence of the specific conditions of GDR early parenting, (b) the traumatic disruptions of unification, and (c) the behavioral models and institutional conditions experienced by West German women. They combine with and add to the above causes.

Among the central issues in this unresolved debate are the juxtaposition of (a) distal versus proximate causes, and (b) values versus rational choice calculations. In this context, "proximate causes" refer to the conditions related to situations and decisions in the immediate context of first birth, while "distal causes" refer to the indirect effects of earlier conditions on childbearing, such as education, careers, and finding a partner. Another important issue in this debate is (c) the question of whether the delay in family formation is more attributable to women's life circumstances and preferences, or is more attributable to those of men. In this chapter, we limit ourselves to looking at the role of partners and spouses from the perspective of women. In another study, we examine the same issues from the perspective of men (Mayer and Schulze 2009, Chap. 5).

The comparison between East and West Germany should allow us to open a special window in regard to several of these issues. Although family policy incentives for marriage and first births were curtailed in East Germany after reunification, child care facilities continued to be much better in East Germany than in West Germany (Trappe 2006). East German women still spend fewer years in schools, continue to have higher rates of labor force participation, and tend to see combining work and motherhood as less problematic than West German women (Statistisches Bundesamt 2006: 523). East German women had been more accustomed to handling economic scarcity and cramped housing conditions, but were exposed to a sudden increase in consumption options after 1989. From the 1970s through to reunification, East German women had a "culture" of early births with or without marriage, and a very low rate of childlessness. This was not least supported by generous family policies (Huinink and Kreyenfeld 2006; Bernardi and Keim 2007; Sørensen and Trappe 1995). Although problems of unemployment also affected West German women, and they too had difficulties securing good jobs after completing their education, these problems were exacerbated for East

German women after reunification (Diewald 2006; Trappe 2004). The East–West comparison allows us, among other things, to study the impact of education on the onset of parenthood within two German sub-societies characterized by different scales and structures of social inequality. East Germany was and is a more homogeneous society with fewer class barriers, while West Germany has a pronounced class structure based on its educational and training systems. As a consequence, we may expect to find smaller educational differentials with regard to the onset of family formation among East Germans than among West Germans. It is unclear whether (and to what extent) the limited role of educational differentials has been maintained since reunification.

This study focuses on East and West German women born in 1971. They are special and representative in a broader sense. The West German women of this cohort made greater advances in attaining higher education and vocational or professional training. Toward the end of the 1980s, they had considerable difficulties in finding apprenticeships, but found it easier to enter the labor market after their apprenticeships (Hillmert and Mayer 2004; Mayer 2004: 204–209). The East German women of the 1971 cohort had mostly completed their vocational training when the economic restructuring started in 1990. They were faced with the risk of job loss and interruptions in their careers, and thus had high rates of participation in retraining and occupational reorientation programs (Trappe 2004; Matthes 2002, 2004). This East German cohort, together with the 1970 cohort, was the first to start the massive delay in age at first birth (Dorbritz 2008: 564; Scheel 2007: 30).

In terms of methods, we use retrospective survey as well as panel data to establish the main explananda, i.e., the age of entering unions and of first birth. We use the quantitative data for singular cases to reconstruct the educational, family, and work trajectories. We use qualitative data from narrative interviews to identify the subjective vocabulary employed when discussing motives for delaying family formation. In presenting and interpreting the material from our narrative interviews, we focus on five general causes or mechanisms that may contribute to the delay in family formation for women: (a) the consequences of lengthening the education and training period, (b) labor market and economic insecurity, (c) value orientations regarding marriage and parenting, (d) problems of compatibility between employment and raising children, and (e) conditions pertaining to a partner or spouse. We also focus on some specific conditions related to education and social class that may enhance or diminish the likelihood of earlier or later entry into parenthood. These are primarily related to the length of time spent investing in education, training, and a career, and the assets the partners bring to a union, which in turn determine their class position.

The chapter is organized in the following sections. In Sect. 4.2 we offer a brief summary of competing theories of the timing of first parenthood. In Sect. 4.3 we provide contextual information on East and West Germany for the women of this cohort. In Sect. 4.4 we describe our data sources and discuss the advantages of combining quantitative and qualitative data and the respective methods of analysis. In Sect. 4.5 we present, as quantitative background, a description of Kaplan–Meier estimates for cohorts born between 1929 and 1971 for three aspects of family formation, non-marital unions, marriages, and first births; as well as Cox regressions for the effects of education on family formation. In Sects. 4.6 and 4.7 we present four case studies each of West German women and East German women. In Sect. 4.8 we summarize the results and offer a preliminary assessment of our experiences with mixed methods.

## 4.2 Theory

In this section we provide a brief review of extant theories on the timing of family formation and the potential effects of social class and education. Our primary goals in this discussion are to guide the selection of topics in our empirical material, and to reflect on our choice of method.

Theories explaining the timing of family formation can be roughly sorted into five categories: (a) life course contingencies and biographical experience, (b) value commitments, (c) rational choice cost–benefit calculations, and (d) institutionally constrained opportunities (Huinink and Konietzka 2007). While these approaches focus primarily on the respondent's characteristics, an additional mode of explanation focuses (e) on the characteristics of the partner or couple.

Theories informed by a life course perspective see the timing of family formation as an outcome contingent on prior lifetime commitments in several life domains. Blossfeld and Huinink (1991) posit an "institutional effect", according to which family formation and parenting are, given fertility control, ruled out as behavioral options while people are in school and training due to an assumed lack of income and time resources. Accordingly, the increasing time spent in education and training (Brückner and Mayer 2005; Jacob 2004) should lead to a delay in the onset of family formation (see also Huinink and Mayer 1995; Kreyenfeld 2006). To this set of potentially delaying life course contingencies we also have to add the transitional phase of career entry leading to a permanent contract. Thus the life course contingency theory would predict delays in the onset of family formation associated with extended durations until the end of the first phase of career entry.

Biographical uncertainties form the second set of explanatory conditions. The experience or expectation of unemployment, instability of jobs and earnings, and residential mobility should lead to delays in entry into marriage and parenthood (Bernardi et al. 2008; Kurz 2008).[1] Entry into marriage and parenthood entails long-lasting commitments which are avoided or delayed, especially if employment and career prospects are unclear. Among our cohort, biographical uncertainties would have been especially acute for the East Germans, although their subjective experiences might have been modified—either enhanced or mitigated—by the fact

[1] An earlier version of a theory connecting biographical turbulence to a decline in fertility was put forth by Birg, Flöthmann and Rester (1991).

that, for East Germans this was more a collective than an individual fate (Huinink and Kreyenfeld 2006).

Changes in value orientations are a third assumed cause of the delay in marriage and parenthood; specifically, the shift towards more individualistic, hedonistic, and consumption-oriented values and lifestyles. Life as a single- or a double-income childless couple allows for much higher levels of consumption than life as parents with one partner earning little or nothing (Brewster and Rindfuss 2000). Value orientations are either assumed to be stable and formed prior to the transition to adulthood, or they are assumed to be dependent on age and life course circumstances. For example, a hedonistic lifestyle might deliver diminishing returns, and thus value orientations in regard to marriage and parenting might change over time. Peer group influences may change over time as friends start to get married or have children. Value orientations, both as permanent and as variable dispositions, differ between East and West Germans. For example, West German women, much more so than East German women, believe that mothers should stay at home while the children are very small (Statistisches Bundesamt 2006: 523).

The fourth set of potential conditions relate to rational decisions based on the situational costs versus the benefits of having children (Becker 1981; Huinink and Konietzka 2007). These costs are higher for highly educated and trained women due to their opportunity costs in career advancement and income loss when they choose to have a family.

Furthermore, institutionally determined opportunities and constraints set the parameters for individual and joint decisions. These apply above all to the provision of childcare and other conditions that tend to foster or hinder the combination of full-time employment and having children. The GDR provided public full-day facilities for toddlers through to school-age children. Although these facilities lost partial support after unification in East Germany, they have continued to function on a much higher level than in West Germany (Büchel and Spieß 2002; Trappe 2004; Huinink and Kreyenfeld 2006).

Finally, whatever the dispositions and characteristics of women may be, family formation also depends on the orientations and life circumstances of men as partners and spouses, and the manner in which men's and women's resources and dispositions are negotiated within those partnerships (Dinklage 2005; Helffrich and Essbach 2004; Kreyenfeld 2006; Kurz 2005; Schmitt 2005).

Huinink and Mayer proposed a theory regarding historical changes and the current relationship between social inequality and family formation:

> The families of origin and the socio-economic backgrounds of husband and wife, as well as the social status of the husband, were important and very visible in the first two [historical] phases of this process. In the first phase, the families of origin tended to be the major actors, at least for the main inheritors. In the second phase, the status of the husband and the parental resources of the wife dominated the family formation process. The question then arises: how does social stratification enter into family formation in its most recent phase?... first, the pattern of family formation for women (timing and events) are regulated to a large extent by women's educational participation, educational inequality, and prospects in the labor market; second, for men, their placement and career in the occupational class structure remains pervasive (Huinink and Mayer 1995: 169).

Huinink and Mayer argued that, in recent periods, traditional norms regarding gender roles and parenthood have mostly lost their regulative force in individual life planning. Direct parental influences on mate selection tend to become marginal and social pressures in regard to the timing and number of children born have diminished markedly in comparison with earlier historical periods. New normative models of parenting become effective, and children are treated more and more like "luxury goods," with the quality of children becoming much more salient than the quantity. The development of "child-centered" families gives rise to self-centered parents or non-parents. The norm of "responsible" parenthood provides many arguments for not having children, or for delaying parenthood. In the "egalitarian" mode of family formation, both men and women have more individualistic attitudes toward partnership and family, depending on their current and anticipated life course conditions and plans.

Under these circumstances, the expectation that fathers will take on the "male breadwinner" role is no longer taken for granted, and it becomes crucial for women to invest an extended period of their lives in education and preparation for a career, or at least a steady source of income. The meaning of women's education has shifted in the sense that women have become more interested in pursuing vocational and professional training than in getting a general education. Having gained equal or perhaps even more than equal access to general education, it is rational for women to take advantage of their investments by seeking skilled jobs with career prospects. Thus more women pursue education, training and a career and this has major consequences for the timing of marriage and childbearing.

As women come to see themselves less as part of a family as a collective actor maximizing the welfare of the family as a whole, a less extreme form of the division of labor within the household makes sense: "Women can pursue their own goals only insofar as they are achieving more equal power within the family... Thus it is in the interest of women to contribute not only a supplementary, but rather an autonomous and equal share of the household income" (Huinink and Mayer 1995: 173).

Because increasing numbers of women are participating in secondary and higher education, and are attaining correspondingly higher-status employment positions, women are now, for the first time, primarily stratified by their own status resources, and to a much lesser extent by the status resources of their parents or their husbands. Thus, women will be increasingly stratified in regard to (a) their educational and skill resources, and, as a consequence, in (b) the temporal transition to marriage and parenthood based on the amount of time spent in education and training, (c) their ideas and perceptions of norms regarding partnership and family, (d) the proportion of their lives they commit to employment, and (e) the economic resources they bring to the household budget.

How then would we expect parental social class and an individual's own education and class position to affect the timing of family formation? Coming from a higher social class may bring with it the experience of a childhood with siblings, and without the deprivation suffered by larger families of less advantaged status. Women who come from a higher social class are more likely to pursue higher education. This should prolong the overall period spent in education and training among women coming from higher social classes. However, because material resources provide greater certainty about goals and their attainability, women who belong to higher classes may experience fewer delays before starting a career than women who come from the lower classes. For women of lower-class backgrounds, committing to a career is a prerequisite for achieving (upper) middle-class status. Thus, delaying marriage and childbirth might be part and parcel of the mobility process, while women of higher-class backgrounds can more easily afford to start a family with the material backing of her (and her partner's) parents.

Women have also changed their position in the marriage market. Whereas in the past, the market position of the husband (and his inherited wealth) determined his position in the marriage queue, for women, that position was largely determined by the relative rank and resources of her parents. In the "egalitarian" mode of family formation, the skills and earning power of women become an asset in the marriage market. Investments in education and career therefore should guide women and trigger a delay in the onset of family formation. We can therefore distinguish prototypically three class-specific strategies of family formation. First, women who have only basic levels of schooling, no training, and a low earning capacity are relatively unattractive on the marriage market, but they often see marriage and motherhood as their main avenue to gain status. That should either result in early marriage and first birth (at times, non-marital), or a delay due to the difficulty in finding an attractive partner. The second strategy is pursued by women with middle levels of schooling and training, e.g., *Realschule* and apprenticeship. They can enter the labor market relatively early, and do not typically have very strong career commitments. Thus, they should marry relatively early and may also have children relatively early. The third strategy of family formation is combining higher education and substantial career investments. This should result in significant delays in marriage. In these contexts, educational homogamy becomes the rule of the game (Blossfeld and Timm 2003) due to the following processes. First, with extended periods spent in education and training, the learning site becomes the favored meeting ground. Second, egalitarian mate selection implies that preference is given to partners who are alike in their orientations and attitudes, and similar in terms of the resources they bring to the union.

In West Germany, and to a lesser extent in East Germany, marriage and/or cohabitation precedes childbirth. Thus, finding the right partner who is able and willing to become a father is a necessary precondition of parenthood. Establishing a secure career is therefore a crucial step not only for women, but even more so for men. This might be problematic in the lower-class strategy, since potential partners might be well out of their education and training period, but still not anchored in a secure career. In the middle-level strategy, both women and men may have the best chances of having completed their training and entry into a career, and to be ready for family formation. The upper-middle-class strategy appears to be the most problematic. Given the length of German university studies, highly qualified women enter the labor market relatively late, and subsequently need some additional years to establish their careers. Since power relations within the partnership or family are maintained either by the lower status of the woman or by an age differential (or both), highly qualified women might have particular difficulties in finding the right partner at the right time.

The effects of parental class background, one's education, as well as the individual's own class situation, operated very differently in the processes of family formation in West and East Germany, both during the division of Germany, as well as in the period immediately following reunification.

Both ideologically and in practice, class differences in the GDR were small. Due to the mass exodus of the self-employed and the upper-middle class, even most university graduates were of working-class origin. Class leveling was further advanced by very selective access to upper-secondary education, which favored those with working-class backgrounds and/or political loyalty. We therefore can expect more cross-class marriage and cohabitation in East Germany. This, in turn, widens considerably the pool of potential marriage partners in East Germany relative to West Germany, and should lead to earlier family formation and parenthood in East Germany.

## 4.3 The East and West German Contexts of Family Formation

Even though after reunification in 1990 both East and West German women were subject to institutional settings and policies that were technically identical, their respective contexts for family formation still differed in major ways. First, many potentially influential differences were fixed before 1989/1990. These differences are above all related to the kind of educational and vocational investments made by young women in the two parts of Germany, as well as to the social norms regarding women's full-time employment and the desirable ages for marriage and childbirth. The East German normative and actual regime of a very early start to family formation—based in part on generous family subsidies and access to housing (Huinink and Wagner 1995; Meyer and Schulze 1992; Kreyenfeld 2006; Scheel 2007; Sørensen and Trappe 1995)—not only set an example for the women of the 1971 cohort, but affected them at least to the extent that they lived in the GDR until about age 19. The age-specific birth rate of this cohort rose from 1986 to 1991, i.e., up to age 20, fell until age 23, and then rose again. The result is an age-specific birth rate with two peaks, one at age 20, and the other at around age 28 (Dorbritz 2008: 564; Scheel 2007: 30). But very different conditions also applied for the period after 1989–1990. On the positive side, East German facilities for early and all-day childcare were still vastly superior to those in West Germany. On the negative side, the disruptions of the privatization shock were felt particularly strongly by women, and especially by young women who had to find employment after vocational training. The Table 4.1 illustrates some of these differences.

**Table 4.1** Educational attainment of women of the birth cohort 1971 in East and West Germany, in percent

| | East Germany | West Germany |
|---|---|---|
| No leaving certificate | 2.4 | 2.4 |
| *Hauptschule* | 3.8 | 26.9 |
| *Polytechnical School (POS)* | 70.7 | 2.6 |
| *Realschule/mittlere Reife* | 4.1 | 33.4 |
| Vocational *Abitur* | 3.2 | 6.3 |
| General Abitur/EOS | 16.7 | 27.1 |

EOS = Erweiterte Oberschule (selective upper secondary school)
*Source* Microcensus 2000–2005. We thank Markus Klein of the Mannheim Center for European Research (MZES) for these calculations

About a quarter of the West German women in this cohort left school at *Hauptschule* level, one-third with *Mittlere Reife,* and another third with vocational or general *Abitur. Hauptschule* is the lower secondary track, and implies 9 or 10 years of schooling. *Mittlere Reife* is the graduating certificate for the *Realschule* after 10 years of schooling, which is the intermediate secondary track. *Abitur* is the graduating certificate of the *Gymnasium* in West Germany after 13 years, or of the *EOS* (*Erweiterte Oberschule*) in East Germany after 12 years of schooling. Almost all East German women had at least a 10th-grade certificate of the Polytechnical Upper Secondary School, and only one-fifth completed the *Abitur*. Their lower rates of participation in upper secondary education meant that, on average, they finished their vocational training and entered the labor market earlier than West German women. At the beginning of the reunification period, about half of the East German women of the 1971 cohort were employed, two-fifths were still in training, 7 % were out of the labor force, and 7 % were unemployed. Among West German women of the same age, 45 % were still in training, one-tenth were still in school, 38 % were working, 5 % were out of the labor force, and 1 % were unemployed. Unemployment among East German women then rose quickly, to about 10 % through the ensuing years.

East German women of the 1971 cohort often could not enter their desired occupation, and had to, on average, spend more periods in full-time training than West German women, before finding a place in the labor market. East German women changed jobs more quickly, and had to change their occupation more frequently (Table 4.2; Mayer and Schulze 2009, Chap. 4).

These data show that the work trajectories of both East and West German women of our cohort were complex and often difficult, but they were considerably more challenging for East German women. The fact that women in East and West Germany had, on average, a second training period (here defined as full-time training of at least 6 months) could be a major condition for delaying motherhood, especially if we take into account the fact that second training courses are often undertaken after several years of employment. However, in regard to subjective perceptions of occupational opportunities, East and West German women are fairly similar.

**Table 4.2** Labor market experiences of East and West German women of the birth cohort 1971

| | East | West |
|---|---|---|
| Could not realize occupational goal, in percent | 53.4 | 49.3 |
| Mean number of training spells | 1.96 | 1.80 |
| Mean number of jobs | 3.62 | 3.64 |
| Mean duration in first job, in months | 45.0 | 50.0 |
| Mean duration in first occupation, in months | 106.0 | 145.0 |
| Ever unemployed, in percent | 59.0 | 37.0 |
| Mean duration unemployed, in months* | 29.7 | 15.0 |
| Perceived bad or very bad occupational opportunities, in percent | | |
| in 1996–1998 | 16.7 | 17.7 |
| in 2000 | 26.1 | 22.2 |
| in 2005 | 22.5 | 28.3 |

*Source* German Life History Study (Mayer and Schulze 2009, Chaps. 2 and 3). Number of training spells, jobs, and occupational changes refers to the time up to age 34
*For those who have been unemployed at least once

If we follow the lives of our sample throughout the available observation period up to age 34, we find that between 5 and 10 % of the West German women were unemployed in any given month, compared with 10–15 % of the East German women, with unemployment rates increasing with age. West German women up to age 34 were twice as likely as East German women to be absent from the labor force (mostly for family reasons). This shows that East German women tried to hold on to full-time employment despite the severe situation on the labor market (Mayer and Schulze 2009, Chap. 3). However, we can also demonstrate that the likelihood of ever having been unemployed was, at 59 %, almost twice as high among East as among West German women; and that the total amount of time spent in unemployment was twice as long among East as among West German women.

From the data of our panel study (see Sect. 4.4), we can also provide an overview of family forms and family values (Table 4.3). In 2005, East German women were less likely to be married and more likely to be cohabiting, but were more likely to be mothers than West German women. An East German woman of this cohort was almost four times as likely to have grown up in a family in which her mother worked full-time until she was 16 years old. In regard to family norms, East German women were more likely to see marriage as an obsolete institution, and to believe that marriage is not necessary for raising children. Unmarried West German women were three times more likely than unmarried East German women to see children as the most important reason to marry. Among unmarried West German women, twice as many said they would like to marry, but had not yet found the right partner. Among those women who had not yet had a child at age 34, West German women were somewhat more likely to say that their partner did not want children and that children are a burden, while East German women were somewhat more likely to say that their job situation was too insecure. While most of the latter differences are not statistically significant, they are part of a consistent pattern.

**Table 4.3** Family forms and family values—East and West German women of the 1971 cohort, in percent

| | East | West |
|---|---|---|
| Ever married at age 34 | 57.4 | 74.7 |
| Cohabiting at age 34 | 16.0 | 5.9 |
| Mother at age 34 | 73.4 | 68.0 |
| Mother always worked until age 16 | 81.9 | 24.5 |
| Marriage is an obsolete institution, agree | 19.1 | 12.4 |
| Marriage is the best way to raise children | 24.1 | 40.0 |
| Would like to marry, but have not yet found the right partner* | 16.6 | 31.5 |
| Would only marry for the sake of children* | 8.4 | 25.0 |
| A toddler suffers when his mother is working | 26.1 | 46.3 |
| Best form of child care when mother is working | | |
| Father | 28.9 | 58.2 |
| Day care | 54.2 | 9.2 |
| Partner does not want children** | 3.7 | 2.9 |
| Job situation too insecure** | 3.3 | 3.9 |
| Children are too much of a burden** | 4.1 | 3.7 |

*Only unmarried respondents
**Women without children, mean value for 1–5 scale with 1 = full agreement, 5 = full disagreement

Norms and perceptions about child care also differ greatly between the two Germanies. While only 26 % of East German women agree with the statement, "a small child suffers when its mother is working", 46 % of West German women agree. When asked to name the best form of child care, if it is not provided by the mother, 54 % of East German women, but only 9 % of West German women picked day care, while 58 % of West German women selected the father.

## 4.4 Data and Methods

The quantitative data used in this study were collected as part of three different surveys of the German Life History Study (Mayer 2008).[2] All surveys of the German Life History Study concentrate on small ranges of birth cohorts in order to capture fine-grained period and cohort effects, and to focus on retrospective event histories in separate life domains, such as residence, family of origin, marital family (including partners and partner's characteristics), education, training, employment, and careers. Events and transitions were recorded forward in time

[2] The data of the German Life History Study (GLHS) is being distributed through the *Zentralarchiv für Empirische Sozialforschung* (GESIS) in Cologne. Basic information and all methods of documentation can be downloaded from www.yale.edu/ciqle, as well as from the web page of the Max Planck Institute for Human Development in Berlin.

and dated monthly. The 1996–1998 data on the East German cohort born 1971 (n = 610) was collected mostly by computer-assisted phone interviews (Matthes 2002; http://www.yale.edu/ciqle/GLHS/index.html). Sampling was based on a regionally stratified random sample drawn from the former GDR central population register. The response rate was 49.5 %. These data include a few cases in which East Germans of the initial sample had moved to West Germany. The East German 1971 cohort was chosen because its members had finished vocational training at the time that the Wall came down, and were thus the first cohort to enter the post-socialist labor market under the new "system." In 1998 and 1999, we collected data for the 1964 and 1971 birth cohorts in West Germany (n = 2,909), again by telephone interview, in cooperation with the Federal Institute for Labor Market Research (IAB) (Hillmert 2004). The response rate was 66.1 %. Both the 1964 and 1971 West German cohorts were of particular interest due to the economic downswing in the 1980s and 1990s, and the presumed effects of international competitive pressures.[3] The 1971 cohort part of that data set is being used in this chapter.[4]

In 2005, we re-interviewed 1,073 of the 1,805 men and women born in 1971 from both the 1996–1998 East German Study and the 1997–1998 West German Study (Matthes 2005) in our own telephone laboratory at the Max Planck Institute for Human Development in Berlin. Field time for the quantitative panel ran from early 2005 to the end of June 2005, and was truncated due to restricted funds.[5] Selectivity, therefore, is a consideration. However, as Table 4.8 shows, panel respondents and panel non-respondents differ only partially in regard to the extent they had ever married or had a first child at the point of the first wave. West German panel respondents have higher rates of marriage and higher rates of childbirth than non-respondents, while for East German men and women, both marriage and birth rates are almost identical for panel and non-panel respondents. Thus, our West German panel selection appears to have delayed family formation to a lesser degree than the non-respondents. It should be noted, moreover, that the estimates of vital demographic rates with the data from the German Life History Study are (given the small sample sizes) astonishingly robust in comparison to corresponding census figures (see Rohwer and Pötter 2003; Scheel 2007: 57–58).[6]

The quantitative data used here come from retrospective surveys, and are therefore subject to recall measurement error. In regard to event-based life history data, the recall error is only a minor fraction of the total survey error (Groves

[3] This interplay of cohort size, labor market conditions, and policy measures was the central focus of our monograph on the latter two cohorts (Hillmert and Mayer 2004).

[4] All the data of these surveys were collected in cooperation with the survey firm "Infas" (Bad Godesberg).

[5] The panel data for the East Germans again includes cases which had moved to West Germany.

[6] Scheel (2007: 58) computed the East German cumulative cohort fertility up to age 34 from the panel data of the German Life History Study (1.18) and the data from official statistics (1.20). The small parameter difference might be accounted for by the West migrants in our East German sample.

**Table 4.4** Qualitative study birth cohort 1971—distribution of cases according to selection criteria

| | East Germany | | West Germany | |
|---|---|---|---|---|
| Education/training | Lower | Higher | Lower | Higher |
| Women | 5(2) | 3(2) | 4(2) | 4(2) |
| Men | 6 | 1 | 4 | 3 |

Education/training: lower apprenticeship with *Mittlere Reife, POS* or *Abitur*; higher: University and *Fachhochschule*. Numbers in brackets refer to the cases selected for this chapter

1989), and depends to a high degree on both the instruments of data collection, and the time and money spent on editing the data. In the German Life History Study, we made huge investments in both of these aspects of data quality (Mayer 2008; Reimer 2005a, b). For example, the 2005 panel study was combined with a precursor methods study in which we developed and tested computer-assisted questionnaires informed by the psychology of autobiographical memory to minimize recall error (Reimer and Matthes 2006).

How did we select the cases for the qualitative study, and how did we select the cases from the available narrative interviews for inclusion in this chapter? Forming the basis for the qualitative sample were 1,073 realized interviews from the panel study. As a first step, we selected a roughly similar number of primary sampling units (PSUs) for West and East Germany. The rationale for selecting according to PSUs was a desire to reduce interviewer transaction cost. Within each region, we then selected north–south and urban–rural PSUs. Within each PSU, we then drew up a list of men and women with lower and higher levels of education and training, roughly similar to the respective population distributions. These addresses constituted the pool of potential respondents. The process of matching the respondent and the interviewer schedules then, finally, determined the actual qualitative sample. The qualitative sample is somewhat biased because we clearly missed cases of very low educational attainment (only *Hauptschule* in the West and fewer than 10 years of schooling in the East). Out of the 30 narrative interviews, there were 16 women interview subjects, and we selected eight cases, each according to the East/West and higher/lower education/training criteria. In addition, our selection reflects the differences in the overall distribution of women with and without children in East and West Germany. For West Germany, we selected three women with no children and one woman with a child, while for East Germany we selected two women with no children and two women with children.[7] Field time for the qualitative study took place mostly in 2005, with some additional interviews conducted in 2004 and 2006. All the names of the respondents in the qualitative case studies were changed, and local references were "blurred." We also excluded or slightly changed other information through which respondents could be too easily identified (Table 4.4).

[7] In our book (Mayer and Schulze 2009) we have documented all cases in detail. This allows us to compare the cases selected here for interpretation with the other half of the women.

In the present study, we are pursuing a modest mixed-methods strategy that combines quantitative and qualitative data. We do not, for example, run parallel analyses on the qualitative and quantitative data in order to test theoretical assumptions about "causes." Nor are we using the quantitative data to test hypotheses derived from the interpretation of the qualitative material, or, conversely, to probe into mechanisms suggested by the narrative evidence through a quantitative analysis. We also did not select our qualitative cases from particular cells produced by prior numerical evidence in causal models. All these are legitimate and potentially fruitful uses of mixed methods. Our aims are, however, more modest. We will use the full quantitative data to establish (a) the explananda for the present study, i.e., ages at union formation and at first birth for the 1971 cohort in comparison to earlier cohorts, and educational differentials in age at first birth; and (b) to document, as in Sect. 4.3, the contexts of family formation processes in the two Germanies. In regard to the selected cases, we are also using the quantitative materials from two surveys (c) for the factual reconstruction of the educational, training, residential, and family histories. This information provides the basis for the individual case descriptions in the text below. Finally, we are using the qualitative materials from the narrative interviews to probe the causes and mechanisms, i.e., the vocabulary of motives (Bernardi and Keim 2007; Bernardi et al. 2008) suggested by theories as outlined in Sect. 4.2.

Our analytic strategy in selecting, presenting, and interpreting the qualitative data was as follows. We first sorted the materials from the narrative interviews case by case according to major substantive areas, such as social background, childhood and schooling, training and transition to labor markets, careers, partnerships and family formation, as well as political orientations. For this chapter, we then concentrated on the material on partnerships and family formation, presenting to the greatest extent possible non-redundant quotes for each case. As a third step, for purposes of interpretation, we looked for the appearance or non-appearance of certain topics, such as the desire for children, perceptions of work-family balance, and the family orientations of partners. Finally, we assessed the consistency of the respective case interpretations of the two authors.

We are, in addition, looking for mechanisms and topics which are not the main focus of the theoretical debate. In this sense, this should be considered a descriptive analysis in the quantitative part, and an exploratory, hypothesis-generating exercise in the qualitative part.

We suggest that the various causes and mechanisms outlined in Sect. 4.2 ideally require different methods of data collection and data analysis. First, it should be noted that both quantified life histories and qualitative biographical self-accounts provide very rich data on the synchronic and diachronic contexts of individual lives. Thus, the "rich context" argument cannot be claimed only by the qualitative side of the theoretical debate. Therefore, we have to look for more subtle differences between the two approaches. Life course contingencies might trigger a delay in family formation without conscious deliberations on the part of the actors. For example, parents or young girls may not make a decision about their age-dependent fertility probabilities when they decide to enter *Gymnasium* at age 10, but this step might lead to a sequence of activities and commitments in education, training, and professional work which can push the potential onset of family formation from the late teens to the early thirties.

Likewise, continuities and discontinuities of working lives are best captured in standardized event histories, but their subjective interpretation as biographical uncertainties involve perceptions and subjective assessments. Therefore, standardized reconstructions of life courses will most likely produce the best data for capturing life course contingencies and complex job trajectories, while biographical uncertainties will more likely reveal themselves in qualitative interviews.

Similarly, while institutional and context conditions can be derived from systematic comparisons between countries and sub-systems—such as East and West Germany before and after reunification—value commitments should manifest themselves more validly in subjective accounts. Moreover, the question of whether fixed value orientations, adaptive value orientations, or short-term rational choice calculations will prevail can best be determined on the basis of extensive qualitative materials. Data on the life courses of partners and spouses are not well covered, even in extensive life history studies like ours, and evidence of the partner's influences, beliefs, and intentions can be better retrieved from subjective accounts. Finally, before we actually attempt to formulate any explanation or draw any conclusions, it is crucial to establish the distribution of the phenomena under study. Therefore, we will use the data from our representative sample surveys to calculate the age-specific rates of cohabitation, marriage, and first birth. We will use the data on standardized life histories for reconstructing the individual temporal patterns of the education-training-career trajectories, and we will use the data from the qualitative interviews to attribute the degree of biographical uncertainty, the value commitments in regard to marriage and children, and the degree to which situation-bound rational choice deliberations prevail.

## 4.5 Entry into Unions and Onset of First Birth of East and West German Women Born 1919–1971

As a descriptive background, we first look at the ages of first cohabitation, first marriage, and first birth among West and East German women. The main purpose of this exercise is to locate our specific cohort in overall cohort development, and to establish our explananda for the 1971 population as a whole. For West Germany, the estimates describe the women born in 1929–1931, 1939–1941, 1949–1951, 1954–1956, 1959–1961, 1964, and 1971; and for East Germany, the estimates describe the women born in 1929–1931, 1939–1941, 1951–1953, 1959–1961, and 1971. As a short cut, we will denote the cohort groups in the following text by their midpoint years.

*Entry into cohabitation and first marriage—West German women* (Table 4.5). The (relatively few) women of the cohort born around 1930 who entered a

**Table 4.5** Medians and quartiles of the age of women at their first union formation in West Germany

| | Birth cohort | 25 % | Median | 75 % |
|---|---|---|---|---|
| First marriage | 1929–1931 | 21.3 | 23.7 | 26.6 |
| | 1939–1941 | 20.4 | 22.2 | 25.0 |
| | 1949–1951 | 19.6 | 21.5 | 24.6 |
| | 1954–1956 | 20.7 | 23.3 | 30.3 |
| | 1959–1961 | 21.5 | 25.6 | / |
| | 1964 | 22.6 | 25.5 | 30.4 |
| | 1971 | 23.8 | 28.3 | / |
| First cohabitation | 1929–1931 | 24.2 | 26.4 | / |
| | 1939–1941 | 21.1 | 25.3 | / |
| | 1949–1951 | 21.1 | 24.7 | / |
| | 1954–1956 | 19.8 | 22.0 | 27.2 |
| | 1959–1961 | 19.3 | 22.3 | / |
| | 1964 | 20.4 | 23.0 | 27.2 |
| | 1971 | 21.1 | 23.8 | 28.6 |
| First marriage or cohabitation | 1929–1931 | 21.3 | 23.7 | 26.5 |
| | 1939–1941 | 20.4 | 22.1 | 24.8 |
| | 1949–1951 | 19.5 | 21.4 | 23.8 |
| | 1954–1956 | 19.7 | 21.5 | 24.8 |
| | 1959–1961 | 19.4 | 22.0 | 25.8 |
| | 1964 | 20.5 | 22.8 | 26.3 |
| | 1971 | 20.9 | 23.6 | 28.1 |

*Source* German life history study, Kaplan–Meier estimates. Slashes indicate that the respective parameters could not be observed at the time of the interview

non-marital union did so relatively late, by age 26, and later than those who married immediately. The age at entry into cohabitation then fell to age 22 for the 1955 cohort, and increased again up to almost age 24 for the 1971 cohort. Age at first marriage was 24 for the 1930 cohort of West German women, declined to almost age 21 for the 1950 cohort, and then rose again to a very high median age of 28 for the 1971 cohort. Overall stability is much higher if we take age at entry to any union—either cohabitation or marriage—as our criterion. Except for the 1930 and the 1971 cohorts, with median union entry ages of 23.7 and 23.6, respectively, all intervening cohorts entered their first union at around age 22.[8]

*Entry into cohabitation and marriage—East German women* (Table 4.6). East German women born between 1930 and 1960 entered their first cohabitation[9] at around age 22. The 1930 cohort had the highest age at 22.6, while the 1940 cohort had the lowest age at 20.9. The age at first cohabitation rose only moderately for

[8] Slashes denote that the respective parameters could not be observed at the time of the interview. These restrictions apply especially to the cohorts born around 1950 and 1960, who were observed up to age 27/28; and the cohorts born around 1955, who were observed up to about age 34/35 (for the survey designs, see also Mayer 2008).

[9] Cohabitation is defined here as sharing a household together for at least one month.

**Table 4.6** Medians and quartiles of the age of women at their first union formation in East Germany

| | Birth cohort | 25 % | Median | 75 % |
|---|---|---|---|---|
| First marriage | 1929–1931 | 20.9 | 22.5 | 25.7 |
| | 1939–1941 | 20.0 | 21.4 | 23.8 |
| | 1951–1953 | 19.8 | 21.1 | 23.6 |
| | 1959–1961 | 20.0 | 21.5 | 24.3 |
| | 1971 | 24.5 | 32.8 | / |
| First cohabitation | 1929–1931 | 20.5 | 22.6 | 28.9 |
| | 1939–1941 | 19.2 | 20.9 | 24.4 |
| | 1951–1953 | 19.4 | 21.3 | 25.1 |
| | 1959–1961 | 19.3 | 21.1 | 24.8 |
| | 1971 | 19.8 | 21.8 | 26.1 |
| First marriage or cohabitation | 1929–1931 | 20.6 | 22.2 | 24.8 |
| | 1939–1941 | 19.8 | 21.3 | 23.5 |
| | 1951–1953 | 19.4 | 20.8 | 22.8 |
| | 1959–1961 | 19.3 | 20.8 | 22.6 |
| | 1971 | 19.8 | 21.7 | 25.5 |

*Source* German life history study, Kaplan–Meier estimates

the 1971 cohort, to 21.8. A quarter of the East German women of the latter cohort started to cohabit by 1991, at the height of the transition turbulence. In this respect, they did not behave differently from earlier generations. Half of that cohort was cohabiting by 1993, and only the third quarter passed through initial cohabitation by 1997 with a further delay of about 2 years (age 26.1). Thus, cohabitation behavior still followed conventional ages for most women of the 1971 cohort. Ages at first marriage among East German women trailed ages at first cohabitation very closely—with about half a year delay—until the cohort born in 1960. We observe a phenomenal increase for the East German 1971 cohort, when the median age at first marriage, 33, rose beyond that of West German women. If we are looking at ages at any union entry, we find a very remarkable degree of stability: all cohorts show median ages between 21 and 22, with a notable increase in the variance for the most recent cohort.

*Transition to first birth—West German women* (Table 4.7). The delay in childbearing among West German women was even more pronounced than the delay in marriage. Women born around 1930 were 26 years old when they had their first child. This age dropped to 24 for the 1940 cohort, and has since risen continuously, reaching age 31 for the 1971 cohort. Kreyenfeld (2006: 19) gives a corresponding estimate of 29.8 for the 1970–1973 cohorts based on microcensus data. To the extent that we are already able to observe their delay in childbearing, given data and time limitations, it is pervasive throughout the cohort, e.g., even the age at which 25 % of women pass the childbirth threshold has shifted upwards, corresponding to the shift in median age.

*Transition to first birth—East German women* (Table 4.7). Until reunification, age at first birth for East German women decreased, from a median age 23.5 for the

**Table 4.7** Medians and quartiles for the age of women at the birth of their first child in West and East Germany

| | Birth cohort | 25 % | Median | 75 % |
|---|---|---|---|---|
| West Germany | 1929–1931 | 22.3 | 25.8 | 29.3 |
| | 1939–1941 | 21.6 | 23.8 | 27.9 |
| | 1949–1951 | 20.9 | 25.0 | 31.0 |
| | 1954–1956 | 22.8 | 27.0 | 34.6 |
| | 1959–1961 | 24.0 | 28.2 | / |
| | 1964 | 24.8 | 28.1 | 34.4 |
| | 1971 | 26.0 | 30.8 | / |
| East Germany | 1929–1931 | 21.2 | 23.5 | 26.3 |
| | 1939–1941 | 20.3 | 22.7 | 25.6 |
| | 1951–1953 | 20.3 | 22.3 | 24.8 |
| | 1959–1961 | 20.5 | 21.8 | 23.8 |
| | 1971 | 23.5 | 27.6 | / |

*Source* German life history study, Kaplan–Meier estimates

1930 cohort to 22 for the 1960 cohort. Not only did age at first birth decrease, the age range at first birth for East German women also tightened (the inter-quartile deviation shrank from 5 to 3 years). In 1989–1990, the period birth rate plunged dramatically, falling almost by half. At that time, there was much speculation about whether East German women would just delay childbearing, or whether they would stop having (or having more) children (Dorbritz 2008; Witte and Wagner 1995; Konietzka and Kreyenfeld 2004; Huinink and Kreyenfeld 2006). For our 1971 cohort, we (so far) observe two tendencies. There was a marked delay in childbearing in the aftermath of reunification. The age at first birth rose by almost 6 years, up to age 27.6. Kreyenfeld (2006: 19) gives a median estimate for the aggregate 1970–1973 cohort of age 27.1, based on microcensus data.

However, while the delay in marriage was highly consequential for childbearing in the West, a delay in marriage in the East did not imply that first childbirth was delayed to the same extent. This is partly due to the high incidence

**Table 4.8** Percentages ever married and having a first child—East and West German women born 1971 at age 27 and at age 34

| | Percentage ever married | Percentage parent |
|---|---|---|
| East Germany | | |
| Survey 1996/97 | 28.5 | 38.8 |
| Survey 1996/97—only panel respondents | 29.3 | 36.4 |
| Panel 2005 | 57.1 | 73.4 |
| West Germany | | |
| Survey 1998 | 45.4 | 33.8 |
| Survey 1998—only panel respondents | 52.2 | 38.8 |
| Panel 2005 | 72.3 | 66.0 |

*Source* German life history study—Kaplan–Meier estimates; the estimates for the 2005 panel include the data for all respondents in the 1998 survey

of non-marital births in the East. In 1989, the proportion of single mothers to all mothers in the GDR was 33 %. During the time after reunification, this share grew rapidly, and reached 57 % in 2003, or 36 % points higher than the corresponding proportion of single mothers in the West (Konietzka and Kreyenfeld 2005: 33).

These cohort comparisons underscore the extraordinary position of the 1971 cohort in regard to delayed family formation. While age at first union entry remained constant in West Germany, and increased only moderately in East Germany, age at marriage rose in both parts of Germany, and dramatically so in the East. Thus, in both parts of Germany, delays in childbearing cannot be due to delays in union formation per se. As a consequence, at about age 27 (the time of our first interview), 45 % of the West German women had married, and 34 % had children; at about the same age, 29 % of the East German women were married and 39 % had a child. At about age 34, at the time of our second interview in 2005, 72 % of the West German women had married and 66 % of them had at least one child, while 57 % of the East German women had married and 73 % had at least one child (Table 4.8).

So why is it that women in both East and West Germany are delaying family formation for so long and why is it that East German women, despite all the disruption resulting from unification, still have children earlier? These are the extraordinary phenomena we want to understand.

A first clue can be found in the distribution of age at first birth according to education (Kreyenfeld 2006). In West Germany, the difference in median age at first birth between women with *Volksschule* certificates and women with higher levels of education rose by almost 10 years between the cohorts born around 1920 and the cohorts born around 1950. It then decreased slightly and stayed fairly stable, at a difference of about 8 years (Fig. 4.1). In contrast, in East Germany for

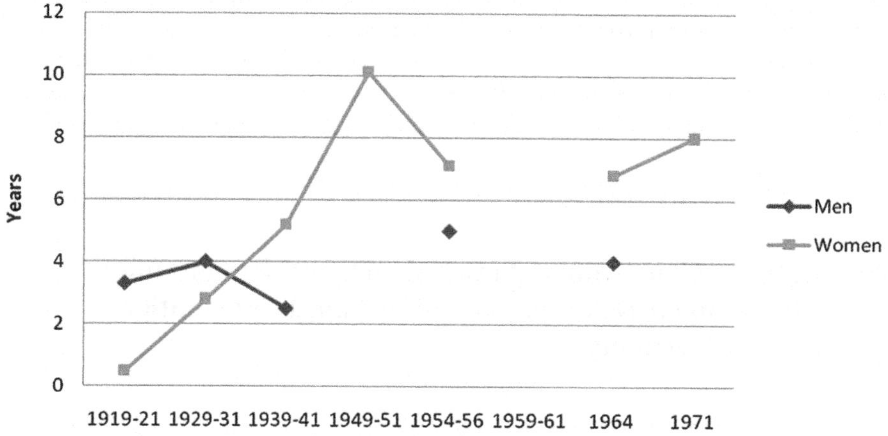


**Fig. 4.1** Difference in median ages at the birth of the first child between the lowest and highest educational categories, West Germany. *Note* In some of the survey years median ages were not observed. For the 1949–1951 and 1959–1961 cohorts this is due to the upper age limit (ca. 28–30 years) of the study

**Fig. 4.2** Difference in median ages at the birth of the first child between the lowest and the highest educational categories, East Germany

the cohorts born between 1930 and 1960, the differences between women of low and high education in the median age at first birth decreased by about 6 years, and almost disappeared. It is only for the 1971 cohort that educational differences had a major impact on age at first birth. This difference shot up by almost 12 years for our most recent cohort (Fig. 4.2). Kreyenfeld (2006) showed that, starting with the 1962–1965 cohorts, and even more so among the 1966–1969 cohorts, the educational differential age at first birth in East Germany had already increased, i.e., those with *Abitur* had already started delaying motherhood before and immediately after reunification.

As Cox models (Appendix Tables A.1 and A.2) show, the changes in educational composition account for almost all of the cohort differences in age at first birth for West German women born between around 1940 and 1964, while only the youngest cohort experienced a further delay in childbearing not accounted for by increasing educational attainment. In contrast, none of the changes in age at first birth in East Germany can be explained by educational composition and its change over time.

## 4.6 Multiple Constraints, Ambivalence and Resignation: Why Women Delay or Abandon Family Formation in West Germany

In the theory section above, we elaborated major reasons why women might delay or abandon family formation: the lengthening and increasing complexity of education, training and career entry, the uncertainties of employment and occupational trajectories, prior value commitments, problems in combining family and work, and the life circumstances/preferences of spouses.

In this section, we present three cases of West German women who are still childless at age 34 in an effort to probe which of the above conditions are manifest and predominant in their articulated vocabulary of motives, and how, in their own view, these motives have changed over time. We will then present a case of earlier childbearing to assess whether and to what extent such delaying factors were absent. Frau Abach, Frau Dr. Grawitz, and Frau Poldinger are examples for the delay in having children due to prolonged periods of training, complex work trajectories, and partner difficulties, which resulted in a reduced or relinquished desire to have children. Frau Kretschmann is an example of a woman who has a lower level of training and who experienced relatively early marriage and childbearing, but who faced similar problems in combining work and family. Frau Abach and Frau Poldinger come from upper–middle-class families. Frau Dr. Grawitz comes from a middle–middle-class family, and Frau Kretschmann from a working-class family. Frau Grawitz and Frau Poldinger have secured upper–middle-class positions by means of their own professional status, but, at least for Frau Grawitz, it is unclear whether her household class position would match her social class of origin (given the class position of her partner). Frau Abach has been sliding downwards on the social scale, because given her own occupational status, she cannot maintain the social class position of her parents.

Moving in together was already too much for him.

Frau Abach: *Abitur*, qualified clerical employee, no career, no current partner, no children.

Frau Abach grew up as the youngest of three children in the South of Germany. Her father was an engineer in a large electric corporation. Her mother worked as a bank clerk, but was at home during her childhood. Her older sister trained as translator; her older brother studied geography and electrical engineering at university, but dropped out of both courses of study. He now works as an employee in a bank. Both siblings are married and have children. After the *Abitur* she had no clear idea of what she wanted to do. She trained for 3 years as a translator, and then switched to a two-year apprenticeship as a hat designer, after which she studied English at university, but quit after a year. At age 26, she moved out of the parental home into her own apartment. She worked as a clerk in the health sector from 1997 to 1999, as a receptionist from 1999 to 2000, and as managerial assistant and editor in the media sector from 2000 to 2002. She then moved to North Germany for a similar position from 2002 to 2003, followed by her current position as a paralegal, which she started upon returning to South Germany, and at which she currently earns 1,700 Euro net per month. Frau Abach has had several relationships, with two of them lasting 2 years, and the most recent one lasting 6 years. In her last relationship, she lived more than 600 km from her partner, but he did not want to move. Shortly after she moved to his city, they ended the relationship. She never cohabited with her partner(s). In her response to the standardized panel interview, she answered that she did not want to have children, but the narrative interview revealed a much more ambivalent attitude. In 1997, she rated her job opportunities as quite good, though less so in 2004. Marriage for her is not important, except if she has children. But, at the same time, "having a good marriage or partnership" ranks very high among her life goals.[10]

> We saw each other and instantly fell for each other. And he said to his mother, who was also there, I would marry this girl on the spot. If he had only done it (laughing)! But then he became afraid. When it got more serious I told him, 'I will move to your city.' But when I got this job I did not want to leave and asked him to join me. But he did not.

Eventually she moved to the city where her boyfriend lived, but into her own apartment.

> Marriage is not so important for me. I did not ask him whether he wanted to marry me. Moving in together would have been the next step. But he didn't want to. That was already too much for him. My goodness, I thought. He dodged the issue and asked whether moving in together would really be good for us, whether we would not get on each other's nerves, and so on... as if it would have been the worst... No, my boyfriend did not want a family. That would have been a responsibility. He is a free spirit, he did not want to commit and he did not want to have children. That was a horrible idea to him... Although he is very good with children and all children love him, he was a hit with my nieces and nephews, but there is nothing I can do if he does not want to have children himself... At the moment I am quite, I am quite happy as it is now. It would be nice if a great man would come along, I would like that. But I am not actively searching either and where should I search? It should just happen again by itself.
>
> To have children is not my wish, and it is not the end of the world if I do not have children within the two next years. But I am sure I could handle it and I am sure I would enjoy being a mother. But if it does not happen, it does not happen. I have many single girlfriends who are also in no rush. So I am not alone in this... Maybe we women must make the decision for ourselves. But I do not want to be a single mom. It is easy to get pregnant, no problem, but I do not want to do it on my own. I want to have a wonderful family like I had myself. It should happen before I get to be 36. Otherwise the age gap will be too big. You probably do not see that well anymore what children need and how they tick. I see this already with my nephews. I have no clue what they are about. Embarrassing!

This long citation expresses Frau Abach's ambivalence well. She loved her boyfriend. She would have liked to have a child, and is confident she would have been a good mother, and would have enjoyed it. But her partner had to agree. Half resigned, she says that she does not need to have a child immediately. But, on the other hand, she does not want to be too old when she has children. Comparing herself with her peers, she concludes that it is not so bad, because her girlfriends have not rushed into parenthood either. And, in apparent resignation, she says, "If it does not happen, it does not happen;" i.e., finding a partner who wants to have a child and getting pregnant.

Thus, for Frau Abach, it is neither a strong commitment to career, nor a desire to spend her money and time on leisure or consumption, which keeps her from having children. The fact that she passed through a sequence of three periods of post-school qualification may have delayed the onset of family formation. It is also noteworthy that she left the parental home when she was 26 years old and always

[10] Response to attitude item in the quantitative study.

lived alone. Also, her move from South to North Germany and back may have delayed settling down. She is furthermore somewhat exceptional in not valuing either marriage or children very highly, while at the same time cherishing her own family and childhood as a high standard to be achieved. While all these aspects might have contributed to a delay, the most obvious reason seems to lie in her relationships. It was her former partner of 6 years who did not even want to cohabit, much less marry and have children. But the reluctance of her partners to commit themselves has a certain counterpart in Frau Abach's own reluctance to settle on a certain occupation and a certain job. Starting a family may also be made difficult given the lofty example provided by her own family.

How has education and social class affected the onset of childbearing in Frau Abach's case? The resources of the parental family allowed her to live at home for a longer time than usual, and to engage in three different episodes of vocational and professional training.

> As a woman in this society, you constantly face hurdles, if you are to advance in a profession.
>
> Dr. Grawitz: medical doctor in specialty training, lives with an unemployed partner.

Frau Dr. Grawitz was born and raised in (West) Berlin. Her mother was born in Czechoslovakia, studied languages, worked as a teacher, and now works as a clerical worker in public administration. Her father died when she was four, and was also a medical doctor. She has a younger brother who only finished *Mittlere Reife*, and completed an apprenticeship as an electronics worker in a big electric firm. He then worked for a big electronics company, but became unemployed when the firm closed down. He now does the same work as a temp. Dr. Grawitz studied medicine and is training to become a specialist in a hospital. She lived with her mother until age 31. Her partner of 9 years is qualified as a metal worker, but worked in a semiskilled job and is currently unemployed. They cohabit but keep two apartments, since he would otherwise lose his unemployment assistance. In early 2005, she told us that she wanted to have two children, and the first within 3 years. Some months later, in the biographical interview, she was more resigned, and only wanted one child. This ambivalence was actually already visible in the standardized interviews, in which she first said she wanted two children, but later fully agreed with the statement, "My occupational future is too uncertain to have children." She also agrees with the items, "Marriage as an institution is obsolete," and, "I do not need a piece of paper to have a stable relationship."

> First I want to finish my training. The hospital where I am working now is okay, but everything is bigger and more anonymous. But I have no choice, because I want to finish my training as quickly as possible. Also in regard to family planning, I am not getting any younger. I will be 35 then, and must decide what I want to do ...I would not want to be 40 when I become a mother ...either it is too late or you still have a chance ...but under no circumstances do I want to interrupt my training now. ...I do not have children yet, but it is still a possibility. But it is difficult. I think as a woman in this society who wants to have a career, you get only stones thrown into your path. I see this with my girlfriends and acquaintances, who are also professionals, who often can't return to their job the way they would have liked to. You really have to think hard about whether you can combine family

and a career ...Originally I wanted to have two children. Now, I think one would be enough. That has changed and it's very sad. But that's just how it is, and it is not right. A friend of mine has a one-year-old daughter. For her it is a bit different because she is a civil servant and has a secure job and she can also work part time if she wants to. But even she is always running into difficulties with finding child care. I think this situation is impossible. It is no surprise that, as I heard on the radio, Berlin women on average do not even want one child. Clearly the basics are somehow wrong... I see my male colleagues, they all go up the ladder, almost none [of the upper hospital ranks] are women. The women just can't manage both... And the men rush ahead and you have no chance to catch up. Or the family suffers. Then I can forget a family... To open a private practice is too risky I would not want to go into debt for that. I want to become an anesthesiologist, and this is pretty difficult... Another possibility would be to go abroad where conditions for child care are better... especially in the Scandinavian countries, where hospitals offer child care. You could not dream of that here. This is totally utopian...I have given up on planning anything, because anyhow you do not know what will come. ...Today it is really not possible anymore to make definite plans. I used to bet that at age 30 you knew that you had a permanent position in a hospital. Now I have just a two-year contract, and then everything is open again...You cannot plan anything. My last contract was for 3 years and that was not bad. I was lucky then. Others had only one-year contracts. In effect, you cannot make plans for the future. You cannot buy real estate, you cannot settle down in a city, that does not work. It is the same for many people, I am not the only one. I will just have to see...For me it was always clear, that I first want to pursue my occupational goals. But even when I was 20, I also wanted to have two children. This has changed. Now I do not think that the absolute way to live a happy life is to raise children. I see this differently now. But I would still somehow like to achieve that. It would be a great experience, even if it is only one child. It is important that you can convey your experiences to the next generation... I see how wonderful that can be among our friends.

My partner has a positive attitude towards children, but... he of course first wants to have a job...And I am lucky that my partner wants to share in parenting. For some time, one could stay at home and then the other. But I think for the first year, you want to enjoy everything, I would want that. But it is a full year and in medicine you can hardly afford it...At the moment, it is up in the air whether we will have children at all. At the moment I don't know. I think I would like to have a child—you might miss something otherwise when you get older. But I do not see it the same way as I did 10 years ago, that it is a must. That it is necessary in order to be fully happy. I do not see it like that any more. Then I could not imagine a future without children. But it was clear that I wanted to pursue my career first. It would be nice to have children, and my partner has the same view, but it does not have to happen at all costs...

Nowadays even 42-year-olds can become mothers. That is not a problem any more. But, no, for me I would not want that. But I see no problem in having a child in my mid-thirties. Of course the clock is ticking, but this is no problem for me... just let it tick. You cannot plan everything. If you then want a child, it might not happen, that's just it. Somehow you will have to live with it. You cannot take it for granted, like when I am 35 years old, I will have my specialty, then we try to have a child and a month later it shall happen, maybe it won't. I will have to see. ...I don't put myself under pressure. That is not how it is with me.

Frau Dr. Grawitz' delay of motherhood reflects the demands and uncertainties of her professional career. She studied medicine and wants to finish her advanced training as quickly as possible. She chose her medical specialty with consideration of its compatibility with raising children, i.e., the possibility of working part time. After finishing her five-year specialty, she wants to start a family, which she cannot do while she is in training due to shift work and long hours. Originally, she wanted to have two children, but has become increasingly pessimistic about it. Now the possibility of having a single child seems to be the best she can hope for. And she already anticipates that having this child will hurt her chances of promotion. Her partner supports her wish for a child, but his long period of unemployment also has negative consequences for their plans, because she anticipates that they will not have enough money to live on once she reduces her work hours. Although the demands and uncertainties of her medical career are clearly an important factor for her, two other aspects come into play. On the one hand, she still upholds the norm that she wants to reduce her working time to care for a small child and cannot see alternatives to that. On the other hand, she does not openly consider the option that her partner could stay at home to care for the child, although he is currently unemployed. Frau Dr. Grawitz wants to combine a professional career and a family, but sees many obstacles in realizing that goal due to the demands and/or uncertainties in her own and her partner's careers. She has already reduced her family goal from two children to one child, and anticipates that even that goal might be unrealistic.

How do social class and education enter as factors in the family formation process for Frau Dr. Grawitz? The class situation of her parents is ambivalent. It is, on the one hand, characterized by the high professional status of her medical doctor father. On the other hand, since he died early, the economic situation of social origin does not match that status level which defined her professional aspirations. Since she has to work to afford her medical studies, she is under greater pressure than her peers might be. A similar class ambivalence is apparent in her choice of partner, who is an unemployed skilled worker. Her experience appears to be a case of upward mobility, aiming to attain her father's social class. Family formation and having children are delayed for the sake of achieving that goal, which assumes priority.

One "demographic" cause for the difficulties of women like Frau Dr. Grawitz may lie in the fact that men at comparable ages with a strong family orientation have already married and become fathers at the time when her career allows her to start a family. Thus, the potential partners left might be those with a weaker family orientation. Another constraint derives from the social norm that professional men can marry less-educated women, but not the other way around. To break this norm and to have a lower-status partner with more time for parenting might be one option for professional women.

> Well, my last boyfriend was totally afraid to have children and to commit to a relationship…
>
> Frau Poldinger: *Abitur* and law degree, works as a lawyer with a large insurance company, strong career orientation, no current partner.

Frau Poldinger was born and grew up in a South German city as the oldest of three children. Her family is part of a left-liberal milieu, and she sees herself as a member of the 1968 generation. Her paternal grandfather was an architect and civil servant. Her maternal grandmother was a tailor whose partner left her alone with two small children. Both parents are architects. Her father worked in several architectural firms before he took a civil servant position in a public planning group. Her mother worked part time in an architectural office and later as a freelancer. After the second child was born, her mother stopped working. Her younger sister is a landscape architect and has one child. Her younger brother earned an *Abitur* and studied painting at the academy. He dropped out after a few semesters and works in a supermarket.

School and university came easily for Frau Poldinger. After the *Abitur,* she worked as an au pair in the United States, spending half a year with a French family and half a year with an American family. The opportunities for law graduates were not promising when she entered university, and she therefore took an international law specialty course in addition to the normal curriculum. She excelled in both areas in her exams, which allowed her to choose her first job: a position with a U.S. law firm specializing in corporate law, where she worked for three and half years. After some years she felt completely overworked and burned out. Also, in order to make the prospect of having a family more realistic, she decided to take a position as a lawyer in a large insurance company, where a family-related reduction to a part-time schedule is possible. Thus, she anticipated the difficulties of combining work and having small children, and planned her career accordingly.

> I very much would like to have a family, and sooner rather than later, because now I am way into my thirties and I think it is about time. But it just has not happened up to now… Otherwise I would have wanted a child some years ago… For instance, when I changed jobs [from a law firm to an insurance company] one reason in the back of my mind was that I could switch to a part-time job in the new firm when I have a family. But I don't have a family. This is due to the fact that I have not found a partner with whom I could have a child… I did have a number of relationships in recent years, but they only lasted for 1 or 2 years, and then we split up again. And in the times in between I was alone…like now… I would very, very much like to have a child or two very soon—we will see how many. Of course, you do not know how it would develop, but I truly would like to have a family. And I also would like to have a man again. But it does not come by itself. Well, my last boyfriend was totally afraid of having children and of committing to a relationship… And before of course I was also younger. So, until 30, nobody thinks of having children. Some of my girlfriends became mothers before 30, but they were all in long-term relationships… And I had a partner who was 30, but having a child was a long-term goal then… and this relationship did not last… If I started a relationship now, then starting a family would not be a distant goal any more… But, as I already also know from my girlfriends, all men are pretty afraid of having children… You must pressure them quite heavily to convince them… Once they are almost 40, a little bit less so… But even then if a woman gets pregnant they panic that they would have to become the breadwinner… At any rate, when men are younger they say they want children, but just not right now. My hunch is the reasons are economic, but also that they are afraid of losing their freedom…Though they say this less openly and might be less conscious of it, in contrast to material reasons.

Frau Poldinger is an example of a successful career woman who would like to have a family in her mid-thirties, but could not (yet) attain this wish because her partners procrastinated or were openly negative about parenting. Now she is looking for a partner with whom she can start a family.

Although her career commitment is high and she can see herself moving up to management level, she is prepared to reduce her workload in order to have small children and has even strategically changed her employer accordingly. Having a child without a partner, like one of our East German interviewees, is not something she can imagine. She wants a "real family." Thus, despite a long training period and heavy early career investments, neither life course contingencies nor biographical uncertainties appear to be the most salient conditions in her case. Her strong value commitments to both career and family are not undermined by how her career unfolds. It is the lack of a partner willing to become a father which produces the observed outcome, although her absorption in work at her first job, and the feeling that "parenting was not an issue before age 30" might have also influenced the partner malaise. Beyond that, as in the cases of Frau Abach and Frau Grawitz, it is a certain ideal of having a family with a close partner relationship, and the desire to devote time to the child which make the realization of this goal difficult.

Frau Poldinger pursues an upper-middle-class position. She wants a high-level career, a husband of similar professional status, and children. Although she has been advancing rapidly in her career and has channeled it in a manner which should make it easier to combine family and career, the main stumbling block is finding the right partner to consolidate her class position.

> We did not really plan to have a child. It was more like, when it happens, it happens.
> Frau Kretschmann: married, one son, saleswoman in a food chain.

Frau Kretschmann was born in an East German town and grew up in a small town in the northern part of West Germany. Her mother worked as an unskilled worker taking care of animals. She died due to alcoholism when Frau Kretschmann was 5 years old. Her father worked in a slaughterhouse and as an employed sales driver working in open markets. At age nine, he married her stepmother, who worked as a saleswoman. She has an older sister who works as a saleswoman, and with whom she lived for a while after the death of her mother, and a younger stepsister. Frau Kretschmann finished *Realschule* and had an apprenticeship as a sales assistant. Her husband is a metalworker. She moved in with him (and his parents) because of trouble with her stepmother when she was 18. At age 19 they moved into their own apartment. She refused to marry until he got himself out of trouble and had finished an apprenticeship (as a scaffold worker). She married at age 25 "for love"[11] and "because she always wanted to marry"[12] and had a son at age 27. In the first interview in 1998, she said she was unhappy because she has no children, and in the second interview she admitted that they did not plan the timing of the child. The child has a genetic defect resulting in partial disability. She is afraid that another child would not be healthy either, and therefore does not want any more children (this is also due to the additional financial burden). She does all

---

[11] Very high agreement with item on attitude scale.

[12] Medium agreement with item on attitude scale.

of the housework and childcare. Frau Kretschmann likes her job and is highly respected in it. She declined an offer to become a store manager, because it would take away time from her family.

> He [her boyfriend and now husband] was quite often unemployed. I had to earn the money for both of us... He did not work much. Then he asked me whether I would marry him. I said no way, if you go on like that, I won't marry you... I said if you learn a trade and get your driver's license, then we can talk about it again... Half a year before he finished his apprenticeship, we married... And I am glad. Now we have been together nine and a half years. Our son is now 5 years old. By now our marriage has endured for so long.
>
> We did not really plan a child, no. We said, when it happens, it happens. ... We had him when I was 27, it was about time. It was okay then. Before we did not want a child, because I saw to it that my husband first finished his training and all, that everything should be a bit more secure. But then we told ourselves, okay, now we have everything as far as that goes, now we can have a child. But one is enough... He [the son] has a muscular [disability]. And then it might be that the next child would have it also, maybe even worse. This risk is too big for me. So for now, I think, it is most important that I can spend a lot of time with him. Because if I have no time for it, why bring a child into the world... We try to do everything for our child, that he should not have any disadvantages. But I say we are happy. Sometimes we are sitting together here and we tell ourselves how well we are doing. What more do we want? And as long as a family is a family, then we can rely on it. Family I find important.

Frau Kretschmann grew up in difficult circumstances and worked hard not only for her own qualifications, but also for her husband's. She is fully employed and takes care of a child with special needs. Marriage and children are taken for granted, but must also be carefully managed. She delayed marrying, but only until her partner (who was unemployed when she met him and a troublemaker) was close to finishing his training. She is clearly the more active partner and thinks that it is up to her to make the family work. Family ranks very high in her life priorities, but work ranks almost as high.

Frau Kretschmann comes from a lower-class background, below that of the secure working class, but attained a middle level of education. She is good at her job and could advance to a supervisory position. But she declined that opportunity in order to be able to work part time and take care of her chronically ill child. She could have started childbearing earlier given her training and employment trajectory, but the marginal position of her husband-to-be delayed not only cohabitation, but marriage and parenthood by about 5 years.

The West German women born in 1971 were, at the time of our interviews, a few years away from the age they considered too late for having a first child. For almost all of them, marriage and parenthood were part of their initial life plans, but these women faced many obstacles in attaining these goals. Particularly those women who had not yet started a family see the possibility of fulfilling those wishes as dependent on their actual or potential partners, and also on perceived difficulties in combining careers and parenting. An important delaying factor seems to be that the actual or former partners tend to shy away from commitments to marriage or children. West German men appear to be the more difficult partners in family formation.[13] It is noteworthy that even those women who have resigned themselves to not finding a suitable partner, and, as a consequence, do not have a strong wish for children, planned their careers very consciously for the eventuality of motherhood (e.g., in regard to part-time options). All the women in our case histories, whether or not they have children, find it difficult to combine work and motherhood. Extended careers and their uncertainties impose additional constraints on professional women.

In general, we can put forward the hypothesis that West German women are subject to an involuntary ambivalence: They want a family and children, but they are negatively constrained by partners and/or careers. Thus, it appears as if it is only partially up to them whether they can realize their goals of family formation. Education and training are crucial determinants for fertility outcomes. Long periods of time spent in education and training, and associated investments in careers, not only seem to delay childbearing; they also—and as a further condition of the former—make it difficult to find a partner who is committed to starting a family. The social class of the woman's parents may help as much as hinder family formation. High parental social class tends to prolong education and training, and appears to set high standards for partners. Improvements in a woman's own high social class position does not seem to accelerate family formation, but rather the opposite (to the extent that it involves heavy investments in education, training, and early career; and to the extent that it is not matched by the class position of the partner).

## 4.7 Family Formation During the Transformation in East Germany: Some Early, Some Late, but Taken for Granted

Family formation for East German women born in 1971 could have been influenced by the norms and behavioral patterns in the GDR, the immediate repercussions of the transformation, and the opportunities and constraints of the new society. Our aim in presenting and interpreting the East German case material is to search for manifestations of all three of these sources of influence. In the traditional GDR pattern, having children was unquestionably taken for granted. In fact, it was a widespread norm to have children in one's early twenties (Kreyenfeld 2006). Two out of the eight East German women in our qualitative sample still followed the pattern of early motherhood. Another characteristic of both the GDR and post-reunification East Germany is the high proportion of unmarried mothers,

[13] The emphasis on male partner reluctance in regard to the delay of family formation of course raises the issue of how reliable women's information about the motives of their partners is. On the one hand, they are corroborated by the cohabitation histories, while on the other hand, we find independent corroboration in the case studies of men (Mayer and Schulze 2009, Chap. 4).

which does not necessarily imply living without a partner. This positive orientation toward parenthood seems to be, in contrast to West Germany, widely shared by East German men (Mayer and Schulze 2009, Chap. 5). The impact of the transformation is manifest in occupational reorientation, residential mobility, as well as job shifts and changes in occupation (see Table 4.2). Finally, West German norms, behavioral patterns and institutional (family policy) provisions could have influenced the family formation patterns of our cohort.

Two of the four following case studies will show the influence of the "Wende" or reunification, with one woman having moved to West Germany for 6 years, and the other living there permanently. Two of our cases show the continuity of former GDR patterns and norms. First, we will present two women with an unbroken GDR mentality, Frau Müller and Frau Pawlak; followed by two women, Frau Magatsch and Frau Mügge, who partially adapted to or are influenced by West German models of female lives.

Frau Pawlak comes from an upper-middle class, bourgeois family, and attained a high level of education and training, as well as a very high-level occupational position. Frau Mügge comes from a middle-class family, but due to reunification did not reach the training levels she aspired to. Frau Müller has a working-class background and current position. Frau Magatsch has a middle-class background due to the professional position of her mother. She herself attained a high educational level and a top professional position.

I was never especially keen on children. It was just normal to have them...
Frau Müller: *POS* (intermediate) schooling, trained nurse, real estate manager.

Frau Müller grew up in two big cities in the North and South of East Germany. She does not know her father, and her mother worked in an accounting office. Her stepfather was a master electrician, and since reunification has worked as a production manager in a large printing firm. She has one stepbrother who is 9 years younger. He started studying computer science, but dropped out. He is now training to become a teacher. She finished school at the intermediate level (*POS*), trained and worked as a nurse. Before the end of her apprenticeship, she had her first son and moved in with the father without marrying him. Her second child, also a boy, was born 5 years later. Shortly after the birth of her second child, she and her partner separated. She blames her first partner for only thinking of himself, of improving his qualifications and leaving them alone while training to become a "Master". In 1999, she retrained for 2 years to become an office clerk because she was looking for work which would allow her to easily combine employment and family. Since then, she has worked in a small company as a facility manager. Also, since 1999, she has lived with her new partner, a lower-level civil servant. They do not want to marry or have children of their own. The older boy has special needs and is enrolled in a special school.

I was never especially keen on children. It was just normal to have them ...It was a mentality during the times of the GDR. Really, it was the first child at age 18 and the second shortly after. We just talked today about what might have been the reasons. Maybe you just had more advantages with children...I don't know. Many of my apprenticeship

> group had the same idea, finishing training [at age 19], then I'll have my first child. There were many who shared this view…

Frau Müller is an example of early childbearing, and how it was part and parcel of the concept of life in the GDR. Having a child early was taken for granted. Frau Müller is also an example of the relative readiness in the GDR, and now in East Germany, to have children and a partner without marrying. Having children is not a hurdle for either separating or finding a new partner. Frau Müller combines work and family and takes responsibility for her children, although she does not feel very close emotional ties with her older son, which she decries. In the question about life goals, she ranks "work that I enjoy" and "a happy partnership" both very high. For Frau Müller, having children early was neither affected by her education and training, nor the class position of her partner's parents or her own. It was something which was normal and expected of women of her age and cohort.

Too bad for the guy, but the next one had to be the father

Frau Pawlak: Career and a child, no partner, executive engineer, living abroad.

Frau Pawlak grew up in or around a large East German town. Her mother completed the *Abitur* and worked as a medical-technical assistant. Her father grew up in the Middle East and migrated to the GDR. He has a university degree and is working as a self-employed media producer. Her maternal grandfather was a high-ranking military officer, her grandmother a homemaker. Her paternal grandfather died before her birth, and was a well-known academic who escaped his home country after a death sentence. She has no siblings. She completed the *Abitur* (*EOS*), and had no definite idea what she wanted to do as an occupation. Her first field of study, business administration, was not her choice, but was assigned to her. From 1990 to 1996, she studied business administration at a technical university and then added an advanced engineering degree in 2000. At age 24 she moved out of the parental home and lived alone in an apartment. She started her career with a permanent contract as a sales manager, but only stayed for a year. After that she worked for another year in the financial sector with a temporary contract. She now works as a high-level utilities manager abroad. At the first interview, in 1997, she said that without reunification she might already have had a child, but would not necessarily be married. She had two partners for short periods of time. Her only child was born in 2003 and is in private day care. Marriage is an obsolete institution for her.[14]

> Well, for me it was more a kind of rational decision, I was 30 and told myself, what do you really want? You have a top job, you have seen a lot of the world, you have had a great time at university, you danced until dawn, and fully enjoyed life, that now the moment has come when it would be good to have a child. Too bad for the guy, but the next one had to be the father, so to speak. It was really that kind of decision: Why should I wait? I did not really want to wait until my dream man appears with whom I want to be together for the rest of my life, and I might be 40 then. And for me it was clear that it would be easy to have a child in this country…My mother said since she was an only child and I was an

[14] Strong agreement with an attitude scale item.

> only child, wouldn't it be nice, if someone in the family would succeed in having more than one child. Well (laughs) it will not be my fault. ...Of course one does not have to overdo it and have one child after the other, but for me it is clear that my child should not remain an only child... I told myself it would be stupid not to use the opportunities here [for easy child care] and wait until I return to Germany, and then I would have the problems [of combining work and family] they all have...No, no...

Frau Pawlak is an example of a single mother who first advanced her professional training and career and then very purposefully became a mother. She did her first and second university degree in rapid sequence, and had no problems entering the labor market. However, she did not immediately find the right job and made two firm shifts up to her present upper managerial position. At age 30 she decided that she might soon be too old to have a child, and so she planned her pregnancy. It was no problem for her that at the beginning of her pregnancy she knew that the father of the child was not going to stay. The optimal conditions for child care (private day care) and her proactive way of combining her job and her duties as a mother (she would bring the baby along to meetings), seems to imply a very positive context for her decision. And she can imagine having a second child under similar circumstances. Living abroad, she has the ability to hire affordable child care. Frau Pawlak is pursuing her goal of motherhood irrespective of a demanding career, living abroad, and not living in a union. She takes it for granted that she will have at least two children, but she also realizes that this is made easier by living in a country with low-wage child care. In comparing her exceptional situation with that of other women in Germany, she sees her situation as ideal and plans to have a second child while she is still abroad. She values marriage very little, but would like to combine having a family with a good partnership. However, this was not a necessary condition, so having a child was a priority for her. Frau Pawlak represents the older GDR mentality, which embraced parenthood as a "natural" part of one's biographical self-conception. What is new is that this mentality now is no longer supported by institutional structures of comprehensive child care and targeted family policies.

Frau Pawlak is influenced by two traditions pushing towards motherhood: the early fertility norms of the former GDR, and, most likely, the strong family norms of the home country of her father, and of the country where she now lives. Her considerable economic resources allow her to have a child, and probably soon two, without assistance from her family, because she can afford personalized child care.

> ...already being planned, the project is in the making.
>
> Frau Mügge: textile merchandiser; lower secondary education (*POS*), apprenticeship as shop decorator, lives with partner in their own house; they have no children.

Frau Mügge grew up in an East German industry town near the Polish border. Her mother was a chemical engineer, her father a Master craftsman. She had a younger brother who died early. Her occupational aspirations were to train as an advertising decorator and then to get further training in advertising. With reunification, the latter option was closed off and she moved to the West to work as a shop decorator in a department store. That move was easier because her then-boyfriend had already moved to the West as well. She could not stay in the firm where she had worked in the East because most workers were laid off. During her time in the West, she obtained further training at an evening trade school in marketing and communication. After six and a half years in the West, she returned to the East, cycled through several jobs, and currently works as a sales promotion manager. For 4 years she lived with a West German, but broke off the relationship because she couldn't stand the West German type of male behavior. At first she was impressed by his polished manners, but after a while she found him too superficial and prone to bragging. Her current East German partner is a construction engineer, and they have rebuilt the small house he inherited. She would marry "if he would ask her," and is trying to get pregnant.

Frau Mügge was all set for the typical very early East German family formation, but the transformation complicated her training aspirations and job trajectory. As a consequence she moved regionally and changed partner. She retained her aspirations to have a family and children and to continue working, but external conditions delayed that process.

> Without reunification my life would have evolved quite differently. Not necessarily better...but much less spectacularly. I would already be married, maybe already divorced again, would have had two children, after some time we would have gotten an apartment and about now we would have gotten the car I applied for when I was 18. We would have a small garden somewhere and go to the Baltic Sea or the Harz for vacations. And most likely I would be still in the same firm where I had started to work; it was unusual to change firms. But I would not have been happier...Reunification came at the right time for me and for me it could have been even earlier. Because then I would have completed my Abitur and gone on to study and would have had a more normal career.
>
> My girlfriends are now all having children. And frankly before age 30, that was not an issue for me... My clear idea was that I would only become a mother when and if the right partner were there, and you have to have some financial security first. And you first have to have good training... and I feel good that I did the jobs I wanted to do, whatever I wanted... so you do not feel sorry if you put less into your career for a while. I would never want to stay at home all day, but I would suspend work for a year and work part time for a while after that. But of course you do not know beforehand whether that is a possibility or not... And I do not think I can do my present job part time. But that is not the important thing, if you have a child, there is always one of us who can take care of the child, especially since my partner has flexible working hours...
>
> It is not that I don't care about marriage, but it is not a high priority. But I wasn't asked yet. I would not ask myself, or ask a man... like women do nowadays ... No, no, but I do want to marry sometime in my life, but it does not matter whether it is now or in a year, but I would not do things like giving the child the name of the father if I was not married, I would never do that. Better to marry then and have the father's name, or the child has my name. Some of my friends did that to have less trouble with changing the names when they marry. But why does one still have to marry? And of course many women have children and are not married; I find no problem in that either... I just find it better if the child has the same name as both parents when it enrolls in school. You feel more like a family. I think it is also nicer for a couple, if you are married and you can say 'my husband' and not 'my boyfriend.' It is more of a commitment. But it is not absolutely necessary... and I am not somebody who says I will leave you if you do not marry me. [How about children?] That is already being planned, just in the making.

Frau Mügge is the example of the East German woman for whom the "Wende" had a considerable impact on her work and family trajectory. She lost her job in 1991, and immediately moved to West Germany. She continued to work in the occupation for which she trained, but since the advanced training she had planned was more complicated, she enrolled in a private evening vocational college. After returning from West Germany after 6 years, she searched for a job without finding a position she liked. After three and a half years as a shop decorator, she switched to a firm where she did some internal training and is working as a merchandiser. Reunification also had quite an impact on her private life. Her then-partner moved to West Germany, and she followed him. One and a half years later they split up and she started a relationship with a West German man. After more than 6 years, she separated from him and returned to East Germany. She returned and found an East German partner. Quite in contrast to her can-do mentality otherwise, she wants her partner to propose marriage. She has an ambivalent attitude towards the connection between children and marriage. More in line with West German models, she anticipates interrupting her employment, and subsequently reducing her working hours, but also anticipates having problems doing so in her current job.

Thus, despite all the disruption in her life course and some uncertainties regarding the future, having children, with or without marriage, is taken for granted. Frau Mügge retains her GDR family ideals, but, like her friends, is willing to delay family formation for a few years. Receiving advanced training and launching a career during the transformation period pushed her to start a family later, but the economic position of her partner actually allows her to actively plan that now.

> ...here in the West ...you always have to choose between a job or children.
>
> *Frau* Magatsch: *Abitur* in the East, studied journalism and works as a television editor in a large Northwest German city, has a partner, does not cohabit.

Frau Magatsch grew up as a single child in a large city in Northeast Germany. Her mother works as a biology and chemistry teacher, and divorced her father when she was 2 years old. Due to her very good grades, she could enter the *EOS*, although this was not unproblematic because her mother was not in the *SED* (Socialist Unity Party of former East Germany) and also not working class.

Since the time she was 14 years old, she wanted to become a journalist (which also made access to the *EOS* difficult) and knew she could do that only if loyal to the party. But even then, this would have been unrealistic in East Germany, because only 20 students per year were allowed to study journalism. After reunification, she was able to realize her career goal and studied journalism at a Bavarian university. She did not like her time at university very much, and found the West German students lacking in knowledge and ambition. After finishing her university degree, she went back to her home town to run a city magazine. Then she applied for work as host of a television show in West Germany. After 2 years she felt burned out, and abhorred the way talk show guests were being publicly exposed. She quit her job and found work as a television journalist in a private production company doing a show about private lifestyles. She was recently promoted to editor-in-chief of the show.

Frau Magatsch left her partner, a West German, after 5 years because she felt that she was too young for a permanent relationship. Her current partner is from her old home town in East Germany and works as a production planning engineer. He is much more eager to start a family than she is. She sees enormous problems in combining her job and having children. He dreams of a house, but she does not want to get into mortgage debt and would be happy just living together.

> … after 5 years I fell in love with somebody else. That was a pity… But I think I just got to know him too early. He would have been the perfect family father. A cool type, knew everything, was tender and loving, very attentive. I got to know him when I was 20 or 21 years old. And then I thought, there has got to be somebody else. Therefore I don't understand people who marry early, or I understand when they separate again soon afterwards. Today you are free to look around. Perhaps you do not always make use of this freedom. What our grandparents did, 50 years of marriage, golden, silver, diamond wedding anniversaries, that is all gone. It is easy to go separate ways. Or the women are really too independent—and that is good. I would not want to have to stay with a man for my whole life just because of the money. Somehow I have the feeling that the current family policies actually keep women in a bit of a state of dependency. That is really a step backwards, which I feel is quite bad.
>
> My current boyfriend is more eager to have children and to build a house than I am. We are of the same age. I think he wants many children and that is something we still have to talk about… and he comes from a small village and is used to living in his own house. I am not so sure about it… I have moved around too often, somehow. I don't know whether I want to be tied down. And not even that, but I don't want to get into so much debt. We don't even live together yet. I was together with my former boyfriend 5 years, then we moved in together, but after 4 weeks I moved out. I don't think I am unable to cohabit, we spend each day together either in his or my apartment. But I just cherish the feeling that I could withdraw if I wanted to… Maybe it is self-centered, I have no idea. But if you have lived alone for such a long time, then you are busy with yourself and have your own rhythm. And at the time I worked for the talk show I was so exhausted when I got home and I wouldn't want to talk to anybody… I could not have been nice to somebody else… Now I am not so stressed out, and find it okay to talk about work, but I cherish putting my feet up and having no obligations, even the obligation to act nicely… As to family policy, then the town here is special… you can hardly find a place in a kindergarten … they pretend to be child friendly, but actually it is a catastrophe.
>
> … here in the West you have to apologize if you bring your small child to a day nursery and if you do not stay at home for 3, 5, or 7 years. …My girlfriends in the East make fun of the West German women who have their first child at age 39. But here it is very, very hard to have children… when I see how badly other women want children, well I don't want to be alone when I am old. I would like if somebody visited me, and put my feet up and brought me a blanket, or whatever… But here one has to make a decision. And I have a hard time deciding against my job. You always have to choose between job or children. You get immediately branded as a career woman even if you just love to work. I really enjoy my job and I am simply afraid that if I stay at home for a while that I won't be able to get my foot back into the door. …
>
> …the family model here in the West is really archaic: … the man goes to work, earns the money, builds a house, buys his wife a second car and brings in the money for the children, and the wife is at home and is happy and thinks she is the greatest, and is happy that she is at home. I find this awful. And the worst is they get a good education before. At

university I met many women who saw it basically as a marriage market and the self-respect of women in the West is so low, it's really ghastly …

Frau Magatsch is acutely aware of the problems of combining family and career. On the one hand, she was socialized in the GDR where the norm was for mothers to be employed. Working mothers received a lot of support and the combination of work and family was not a big issue. On the other hand, Frau Magatsch now works in a large West German city and sees the problems women in her environment have as especially critical. She openly criticizes this pressure to choose between a career and children. She is delaying having children, because she does not want to give up her job. This brings her into conflict with her East German partner, who is eager to start a family. Thus, Frau Magatsch is really caught between East and West. She upholds the GDR norms of combining family and work, but sees no way to put this into practice in the West. She is a good example of a woman with a very strong career orientation that brings her into a real dilemma over childbearing. On the one hand, she has the East German orientation that a child is a normal part of one's life, but she finds this orientation difficult with the conditions she finds herself at as a career woman in West Germany. She is as goal-oriented in her professional life as she is cautious in regards to family formation and childbearing.

The women born in 1971 in East Germany were, at the time of our interviews, even closer than West German women to the latest age at which they think they should have a child, because the norm of early pregnancy is still very strong in the East. Having children is taken for granted, even more so than marriage. The opportunities which opened up after reunification for obtaining new qualifications and reorienting occupational pathways have partly delayed having children, but it is almost never an either/or. The uncertainties and turbulence of the labor market make family life more complicated, and sometimes result in divorce and separation, but do not generally deter family formation. The male partners similarly support such a pattern. Never did we hear that a partner did not want to become a father or would not take his parenting role seriously even after a split-up. After 1989, educational levels strongly differentiated the onset of childbearing (Kreyenfeld 2006). Women with apprenticeships in service vocations either followed the early childbearing pattern of the old GDR, or they delayed having children, but stuck to the goal. Women with academic careers either gave up or delayed childbearing, or had children even without having a partner.

## 4.8 Summary and Conclusions

In this chapter, we attempted to understand the mechanisms underlying the delay of family formation among East and West German women. In particular, we focused on the birth cohort of 1971, who faced the turbulence of the post-reunification transformation in the East, and problems entering the labor market in the

West. In the field of demography, our contribution is methodologically innovative, because we were able to draw on both quantitative and qualitative data for identical respondents from the German Life History Study.

The starting point of our investigation was a series of Kaplan–Meier estimates for the transition to cohabitation, marriage, and first birth. In a historical comparison, the 1971 birth cohorts in both East and West show particularly high median ages at first birth. West German women were, on average, 28.3 years old when they first married, and 30.8 years old when they had their first child, while East German women were 32.8 years old when they first married, and 27.6 years old when they had their first child. West German women of different educational levels differ in the onset of childbearing by eight to 10 years, and the overall delay in childbearing can be explained by the changing educational composition, with the exception of the even greater delay among the most recent cohort. In contrast, East German women showed a trend of ever smaller differences in ages at first birth, until a rapidly increasing differential emerged for the most recent cohort. The analysis of the development of West and East German educational differentials provides a longer-term addition to the study by Kreyenfeld (2006), which concentrated on the 1960s and 1970s cohorts.

We focused on five possible causes for the delay in family formation for women: (a) consequences of the lengthening of the education and training period, (b) labor market and economic insecurity, (c) value orientations in regard to marriage and parenting, (d) problems of compatibility between career and raising children, and (e) conditions pertaining to a partner or spouse. We used the case studies reconstructed from both the standardized responses and the narrative interviews to ascertain the presence or absence of these conditions for each case, and to assess their intra-individual weight. Thus, the purpose of this analysis is not to establish causal weights in the overall population, but instead to reconstruct the definition of the situation, the vocabulary of motives, and the logic these women follow when describing themselves in the process of family formation.

Our combination of standardized and qualitative material provides evidence for mechanisms in the delay of family formation that in part turned out to be quite different between West and East German women. For both East and West German women, combining work and family is difficult, but this more strongly deters West German women from having children. Likewise, early relationships with partners and early non-marital unions rarely lead to permanent partnerships in both parts of Germany, but again such complications keep West German women from having children more frequently than East German women. Not being married or even not having a long-term partner at all does not prevent East German women from becoming mothers.

For West German women, the delay in family formation appears to be almost predetermined, because a number of conditions must be simultaneously present, thus making it quite difficult to establish a family. For almost all women, living with a partner and having children are important life goals. Thus, it is clearly not the absence of family value orientations that prevent women from having children or cause them to have children very late. The phase prior to integration into the labor market with a relatively stable job is prolonged, due to both choice and constraints. This is due less to the fact that more women complete the *Abitur* than to the fact that more women at almost all educational levels have more than one period of training, e.g., by changing subjects at university or adding further training after an initial apprenticeship. Especially for highly qualified women, the uncertainties of employment play an important role. In addition, the practical problems of combining work and having small children are seen as almost insurmountable. Within this context, West German women actively plan their careers to make them compatible with having children by anticipating family leave and a reduction in working hours. However, probably the strongest reason for the delay in family formation seems to be the reluctance of male partners to commit themselves as fathers, and/or problems of employment security and career uncertainties of the male partners. Facing these problems of male ambivalence and lack of support in the work sphere, West German women delay the first birth, reduce the number of children they would like to have from two to one, or abandon their motherhood goals altogether. Interestingly enough, West German women do not want to rely on men as full breadwinners, but anticipate that they will contribute a large share to the family budget.

As socio-demographic data show, East German women born in 1971 on average not only have their first child earlier and marry later than West German women, but also have a much higher occurrence of single motherhood. As regards the mechanisms underlying the processes of delayed family formation, we need to answer two questions: First, do these mechanisms differ between East and West Germany? And, second, to the extent that the mechanisms are similar, do they have the same meaning and impact in the East as in the West? In terms of value orientations, we see not only very clear differences, but also differences in the way they operate. While for West German women, a conscious and fairly positive evaluation of having children seems to be a prerequisite for the decision to become a mother, no such value judgment seems to be necessary in the East. Having children and having children early was taken for granted in the former GDR, and was still very influential for our cohort. A person does not need to like children especially in order to have children. Although quite a few of our West German women do not see marriage as an absolute must, they still see being married as a desirable goal, and ideally a precondition for having children. For East German women, marriage is clearly not a precondition to the same extent. Also important in the value sphere is the strong emphasis East German women place on being fully employed, while West German women all want to suspend and reduce working hours to be a good mother. However, East German women also see this goal as less of a hindrance to motherhood than West German women.

Given the turbulence in their occupational trajectories resulting from the destruction and rebuilding of the GDR economy, East German women should have been much more affected by the mechanisms we study. East German women of this cohort were ready earlier for family formation because they finished their apprenticeships earlier and were much less often enrolled in upper-secondary and higher education. But the interruption of qualification trajectories and stable employment brought about by the "Wende" by far outweighed this earlier potential for family formation. Less than a quarter of East German women followed the former pattern of early motherhood/marriage and then had to work through marital relationships/parenthood under these difficult circumstances. The majority engaged in time-consuming re-qualification, job, and residential changes, and delayed having their first child. In either case, they still took childbearing for granted as a part of their life, irrespective of job changes and employment insecurity.

Our East German women also ideally want to have children with a partner, but the idea of changing partners or being a single mother is much less of a deterrent for them. They cannot imagine being financially dependent on their partner. But the major difference to the West German women is that they are not subject to widespread male resistance to fatherhood. This is all the more noteworthy, since the occupational lives of these partners are not less, but more turbulent than those of their West German counterparts. Their problems on the labor market make family life more complicated, and sometimes result in divorce and separation, but do not generally deter East Germans from starting a family.

Finally, we want to draw preliminary conclusions from our experiences in combining quantitative and qualitative methods. The first advantage we would like to point out is that we were able to reconstruct the education, work, family, and residential histories on the basis of both the monthly calendars of the quantitative survey and the narrative interviews. In many cases, neither the qualitative nor the quantitative data alone would have been sufficient to give an adequate description of these histories. While the drawback of the qualitative materials in this regard is that they are often incomplete as to particular episodes and their dating, the drawback of the quantitative material is that sometimes necessary contextual information is missing, or that standardized responses might be misleading. Second, we were fortunate in being able to select the respondents of our qualitative study as a stratified subset from our overall panel survey. We thus covered the whole range of timing of first birth and of prior union formation. In this chapter, for space reasons, we primarily looked at six cases with delays and two counter examples. Ideally one would look at the whole range (as we did in Mayer and Schulze 2009). Furthermore, on the basis of a qualitative sample alone, there is always the strong temptation to draw erroneous conclusions concerning the distribution of the dependent variable. This temptation is even stronger in regard to the relative weights of explanatory factors. Conversely, by relying exclusively on the quantitative material, we would not have seen or sufficiently appreciated the particular role of men in the family formation process, and the major differences in how mechanisms operate in the East as compared to the West despite similar outcomes. The qualitative accounts were clearly superior, both in reflecting the ambivalence of attitudes toward children, and of changes in preferences over time. But at this stage of our work, it would be premature to make a final assessment of the deficits of either method, and the advantages of triangulation.

**Acknowledgments** The collection of the data used in this study was funded by the Max Planck Society for the Advancement of Science through the Max Planck Institute for Human Development (Center for Sociology and the Study of the Life Course), Berlin, the *Deutsche Forschungsgemeinschaft* and the European Social Fund. Data analysis was supported by Yale University through the Center for Research on Inequalities and the Life Course (CIQLE). We are especially indebted to Britta Matthes for her inputs in the data collections of the 1997 study of the East German 1971 cohort and the 2005 panel study, as well as to Karola Rockmann and Ralf Künster for their help in computer-assisted data collection and data organization. We would also like to thank Jörg Luedicke (CIQLE) for his assistance in computing the Kaplan–Meier estimates and Cox models, as well as other data analysis, Chelsea Rhodes (CIQLE) and Lucas Wiesendang (CIQLE) for editing the manuscript, Miriam Hils-Cosgrove for further language editing, Natalie Nitsche (CIQLE) and Jana Schmidt (BIS) for constructing the case histories from the survey data, and Anja Wilbrandt (BIS) for editing parts of the narrative interviews. The Mannheim Center for European Social Research at the University of Mannheim provided generous support during a sabbatical stay of the first author, and in providing one-week workplace for the second author. Critical comments by the editors and Johannes Huinink greatly helped us in revising the chapter.

# Appendix

See Tables A.1, A.2.

**Table A.1** Transitions into parenthood—Cox proportional hazard models for the birth of the first child—West Germany, women

| | Model 1 | | Model 2 | |
|---|---|---|---|---|
| | Exb(B) | Sig. | Exb(B) | Sig. |
| *Birth cohort* | | | | |
| 1929–1931 | Ref. | 0.000 | Ref. | 0.000 |
| 1939–1941 | 1.182 | 0.035 | 1.262 | 0.003 |
| 1949–1951 | 0.969 | 0.702 | 1.135 | 0.122 |
| 1954–1956 | 0.697 | 0.000 | 0.928 | 0.338 |
| 1959–1961 | 0.607 | 0.000 | 0.855 | 0.070 |
| 1964 | 0.631 | 0.000 | 0.919 | 0.259 |
| 1971 | 0.441 | 0.000 | 0.702 | 0.000 |
| *Education* | | | | |
| No degree, Hauptschule | | | Ref. | 0.000 |
| Realschule, Mittlere Reife | | | 0.689 | 0.000 |
| Abitur, Hochschulreife | | | 0.340 | 0.000 |

**Table A.2** Transitions into parenthood—Cox proportional hazard models for the birth of the first child—East Germany, women

| | Model 1 | | Model 2 | |
|---|---|---|---|---|
| | Exb(B) | Sig. | Exb(B) | Sig. |
| *Birth cohort* | | | | |
| 1929–1931 | Ref. | 0.000 | Ref. | 0.000 |
| 1939–1941 | 1.203 | 0.102 | 1.226 | 0.072 |
| 1951–1953 | 1.350 | 0.008 | 1.386 | 0.011 |
| 1959–1961 | 1.494 | 0.000 | 1.518 | 0.002 |
| 1971 | 0.475 | 0.000 | 0.507 | 0.000 |
| *Education* | | | | |
| No degree, Hauptschule | | | Ref. | 0.000 |
| Realschule, Mittlere Reife | | | 1.053 | 0.612 |
| Abitur, Hochschulreife | | | 0.663 | 0.002 |

## References


Becker, G. (1981). *A treatise on the family*. Cambridge: Harvard University Press.

Bernardi, L., & Keim, S. (2007). Anfang dreißig und noch kinderlos? Lebenswege und Familienmodelle berufstätiger Frauen aus Ost- und Westdeutschland. In D. Konietzka & M. Kreyenfeld (Eds.), *Leben ohne Kinder: Kinderlosigkeit in Deutschland* (pp. 317–334). Heidelberg: VS Verlag für Sozialwissenschaften.

Bernardi, L., Klärner, A., & von der Lippe, H. (2008). Job insecurity and the timing of parenthood: A comparison between Eastern and Western Germany. *European Journal of Population, 24*(3), 287–313 .

Birg, H., Flöthmann, E. J., & Rester, I. (1991). *Biographische Theorie der demographischen Reproduktion*. Frankfurt: Campus.

Blossfeld, H.-P., & Huinink, J. (1991). Human capital investments or norms of role transition? How women's schooling and career affect the process of family formation. *American Journal of Sociology, 97*(1), 143–168.

Blossfeld, H.-P., & Timm, A. (Eds.). (2003). *Who marries whom? Educational systems as marriage markets in modern societies*. Dordrecht: Kluwer Academic.

Brückner, H., & Mayer, K. U. (2005). The structure of the life course: Standardized? Individualized? Differentiated? In R. Macmillan (Ed.), *Advances in life course research* (Vol. 9, pp. 27–54). Oxford: Elsevier.

Brewster, K., & Rindfuss, R. (2000). Fertility and women's employment in industrialized nations. *Annual Review of Sociology, 26*(1), 271–296.

Büchel, F., & Spieß, K. (2002). *Form der Kinderbetreuung und Arbeitsmarktverhalten von Müttern in West- und Ostdeutschland. Bundesministerium für Familie, Senioren, Frauen und Jugend*. Stuttgart: Kohlhammer.

Diewald, M. (2006). The quest for double transformation: Trends of flexibilization in the labor markets of East and West Germany. In M. Diewald, A. Goedicke, & K. U. Mayer (Eds.), *After the fall of the wall: Life courses in the transformation of East Germany* (pp. 269–292). Stanford: Stanford University Press.

Dinklage, M. (2005). *Der Zeugungsstreik. Warum die Kinderfrage Männersache ist*. München: Diana.

Dorbritz, J. (2008). Germany: Family diversity with low actual and desired fertility. In T. Frejka, T. Sobotka, J.M. Hoem & L. Toulemon (Eds.), *Childbearing trends and policies in Europe. Demographic Research, Special Collection* (Vol. 7(19), pp. 557–598).

Groves, R. M. (1989). *Survey errors and survey costs*. New York: Wiley.

Helffrich, C., & Essbach, W. (2004). Männer leben—Studien zu Lebensläufen und Familienplanung von Männern. Resource document. Freiburg University. http://www.maennerleben.de.

Hillmert, S. (2004). Die Westdeutsche Lebensverlaufsstudie, Kohorten 1964 und 1971: Projekt, Datenerhebung und Edition. In S. Hillmert & K.U. Mayer (Eds.), *Geboren 1964 und 1971. Neuere Untersuchungen zu Ausbildungs-und Berufschancen in Westdeutschland* (pp. 215–230). Wiesbaden: VS Verlag für Sozialwissenschaften.

Hillmert, S., & Mayer, K.U. (2004). *Geboren 1964 und 1971. Neuere Untersuchungen zu Ausbildungs- und Berufschancen in Westdeutschland*. Wiesbaden: VS-Verlag für Sozialwissenschaften.

Huinink, J., & Konietzka, D. (2007). *Familiensoziologie*. Frankfurt: Campus.

Huinink, J., & Kreyenfeld, M. (2006). Family formation in times of abrupt social and economic change. In M. Diewald, A. Goedicke, & K. U. Mayer (Eds.), *After the fall of the wall: Life courses in the transformation of East Germany* (pp. 170–190). Stanford: Stanford University Press.

Huinink, J., & Mayer, K. U. (1995). Gender, social inequality, and family formation in West Germany. In K. O. Mason & A.-M. Jensen (Eds.), *Gender and family change in industrialized countries* (pp. 168–199). Oxford: Clarendon Press.

Huinink, J., & Wagner, M. (1995). Partnerschaft, Ehe und Familie in der DDR. In J. Huinink & K.U. Mayer (Eds.), *Kollektiv und Eigensinn. Lebensverläufe in der DDR und danach* (pp. 145–188). Berlin: Akademie.

Jacob, M. (2004). *Mehrfachausbildungen in Deutschland: Karriere, Collage, Kompensation?*. Wiesbaden: VS Verlag für Sozialwissenschaften.

Konietzka, D., & Kreyenfeld, M. (2004). Angleichung oder Verfestigung von Differenzen? Geburtenentwicklung und Familienformen in Ost- und Westdeutschland. *Berliner Debatte Initial, 15*, 26–41.

Konietzka, D., & Kreyenfeld, M. (2005). Nichteheliche Mutterschaft und soziale Ungleichheit. Zur sozioökonomischen Differenzierung der Familienformen in Ost- und Westdeutschland. *Kölner Zeitschrift für Soziologie und Sozialpsychologie, 57*(1), 32–61.

Kreyenfeld, M. (2006). Der Einfluss der 'Wende' auf bildungsspezifische Fertilitätsunterschiede in Ostdeutschland. MPIDR WP 2006-025. Rostock: Max Planck Institute for Demographic Research.

Kurz, K. (2005). Die Familiengründung von Männern im Partnerschaftskontext. Eine Längsschnittanalyse zur Wirkung von Arbeitsmarktunsicherheiten. In A. Tölke & K. Hank (Eds.), *Männer—Das 'vernachlässigte' Geschlecht in der Familienforschung*. Sonderheft 4 der Zeitschrift für Familienforschung (pp. 178–197). Wiesbaden: VS-Verlag für Sozialwissenschaften.

Kurz, K. (2008). *Beschäftigungsunsicherheiten und langfristige Bindungen*. Wiesbaden: VS-Verlag für Sozialwissenschaften.

Lesthaeghe, R., & Surkyn, J. (1988). Cultural dynamics and economic theories of fertility change. *Population and Development Review, 14*(1), 1–45.

Matthes, B. (2002). Brücken und Stolpersteine auf dem Weg ins Erwerbsleben. Die Folgen der Transformation für den Erwerbseinstieg ostdeutscher Jugendlicher. Fachbereich Politik- u. Sozialwissenschaft. *Dissertation*. Berlin: Freie Universität Berlin. http://www.diss.fu-berlin.de/2003/84.

Matthes, B. (2004). Der Erwerbseinstieg in Zeiten des gesellschaftlichen Wandels. Ost- und westdeutsche Jugendliche auf dem Weg ins Erwerbsleben. In S. Hillmert & K.U. Mayer (Eds.). *Geboren 1964 und 1971. Neuere Untersuchungen zu Ausbildungs- und Berufschancen in Westdeutschland* (pp. 173–199). Wiesbaden: VS-Verlag für Sozialwissenschaften.

Matthes, B. (2005). *Dokumentation der Studie Frühe Karrieren und Familiengründung: Lebensverläufe der Geburtskohorte 1971 in Ost- und Westdeutschland (LV-Panel 71)*. Berlin: Max-Planck-Institut für Bildungsforschung.

Mayer, K.U. (2004). Unordnung und frühes Leid? Bildungs-und Berufsverläufe in den 1980er und 1990er Jahren. In S. Hillmert & K.U. Mayer (Eds.). *Geboren 1964 und 1971. Neuere Untersuchungen zu Ausbildungs- und Berufschancen in Westdeutschland* (pp. 202–213). Wiesbaden: VS-Verlag für Sozialwissenschaften.

Mayer, K.U. (2008). Retrospective longitudinal research: The German life history study. In S. Menard (Ed.), *Handbook of longitudinal research* (pp. 85–106). Amsterdam: Elsevier.

Meyer, S., & Schulze, E. (1992). Familie im Umbruch. Zur Lage der Familien in der ehem. DDR. *Schriftenreihe des Bundesministeriums für Familie und Senioren*. Bd. 7. Stuttgart.

Mayer, K.U., & Schulze, E. (2009). *Die Wendegeneration. Lebensverlaüfe des Geburtsjahrgangs 1971*. Frankfurt: Campus.

Reimer, M. (2005a). *Autobiografisches Gedächtnis und retrospektive Datenerhebung. Die Rekonstruktion und Validität von Lebensverläufen*. Berlin: Max-Planck-Institut für Bildungsforschung: Studien und Berichte 70.

Reimer, M. (2005b). *Collecting event history data about work careers retrospectively: Mistakes that occur and ways to prevent them*. Berlin: Max Planck Institute for Human Development.

Reimer, M., & Matthes, B. (2006). Collecting event histories with true tales: Techniques to improve autobiographical recall problems in standardized interviews. *Quality and Quantity, 41*(5), 711–735.

Rohwer, G., & Pötter, U. (2003). *Demography of Germany: Concepts, data and methods*. Version 3. Resource document. University of Bochum. http://www.stat.ruhr-uni-bochum.de/pub/eha/ddem.pdf.

Scheel, M. (2007). *Der Übergang zum ersten Kind in Polen und Ostdeutschland*. Universität Rostock: Diplomarbeit.

Schmitt, C. (2005). Kinderlosigkeit bei Männern. Geschlechtsspezifische Determinanten ausbleibender Elternschaft. In A. Tölke & K. Hank (Eds.). *Männer—Das vernachlässigte Geschlecht in der Familienforschung*. Sonderheft 4 der Zeitschrift für Familienforschung (pp. 18–43). Wiesbaden: VS-Verlag für Sozialwissenschaften.

Sørensen, A., & Trappe, H. (1995). Frauen und Männer: Gleichberechtigung-Gleichstellung-Gleichheit? In K. Huinink, K.U. Mayer, M. Diewald, H. Solga, A. Sørensen & H. Trappe (Eds.). *Kollektiv und Eigensinn. Lebensverläufe in der DDR und danach* (pp. 189–222). Berlin: Akademie.

Statistisches Bundesamt. (2006). *Datenreport 2006*. Bonn: Bundeszentrale für Politische Bildung.

Tivig, T., & Hetze, P. (2007). *Deutschland im Demografischen Wandel*. Rostock: Rostocker Zentrum zur Erforschung des Demografischen Wandels.

Trappe, H. (2004). Chancen West, Chancen Ost—Frauen und Männer des Geburtsjahrgangs 1971 im Vergleich. In S. Hillmert & K.U. Mayer (Eds.), *Geboren 1964 und 1971. Neuere Untersuchungen zu Ausbildungs-und Berufschancen in Westdeutschland* (pp. 133–153). Wiesbaden: VS-Verlag für Sozialwissenschaften.

Trappe, H. (2006). Lost in transformation? Disparities of gender and age. In M. Diewald, A. Goedicke, & K. U. Mayer (Eds.), *After the fall of the wall: Life courses in the transformation of East Germany* (pp. 116–139). Stanford: Stanford University Press.

Witte, J. C., & Wagner, G. G. (1995). Declining fertility in East Germany after unification: A demographic response to socioeconomic change. *Population and Development Review, 21*(2), 38–48.

# Chapter 5
# De-Standardisation or Changing Life Course Patterns? Transition to Adulthood from a Demographic Perspective

**Johannes Huinink**

## 5.1 Introduction

The transition to adulthood for both men and women has undergone considerable changes in European countries since the mid-1960s. Participation rates in higher education have steadily increased. Women have intensified their labour force participation. Traditional family related institutions have less of an impact on early life courses. Marriage and family formation are delayed. Non-marital and childless forms of living arrangements have stepped in to fill the gap. In general, highly institutionalised traditional pathways to social and economic independence have lost relevance for young men and women. They have changed their behaviour to profit from the opportunities the welfare state institutions and the economic market have to offer. At the same time, they have to follow the rules of these institutions to be successful. The need for individual autonomy and flexibility has increased.

In the social sciences, it seems to be taken for granted that an all-embracing de-standardisation of patterns of transition to adulthood, and of individual life courses in general, have been the consequences of these developments. It means that individual life courses become more diverse, that the variance of the age distribution at particular events increases, and that the heterogeneity of life states among members of certain age groups amplifies. Indeed, this view is widely supported by empirical evidence. Findings of studies in which this process has been investigated show a profound and constant trend of a dilution of formerly highly structured passages of the life course in nearly all European countries (Billari and Wilson 2001; Corijn and Klijzing 2001; Elzinga and Liefbroer 2007). Between different

J. Huinink (✉)
Institute for Empirical and Applied Sociology (EMPAS), University of Bremen, Bremen, Germany
e-mail: huinink@empas.uni-bremen.de

G. Neyer et al. (eds.), *The Demography of Europe*,

DOI: 10.1007/978-90-481-8978-6_5,

countries, considerable diversity can, however, be observed with regard to both the onset and the extent of this change (Billari 2004).

In the following, I will argue that this comprehensive view of de-standardisation falls short of the logic of processes of social change. There are good theoretical and empirical reasons to expect that patterns of transition to adulthood just change and new standardised patterns of this status passage in the life course arise. Reverse trends to re-standardisation can emerge. In many cases, de-standardisation can just be a transient phenomenon, occurring during periods of pattern change and resulting in new, but again highly structured patterns of life course trajectories. Therefore, the general impression of an increasing diversity of individual life courses and transition to adulthood might be "biased".

In this chapter, I discuss this issue from a conceptual, theoretical, and empirical perspective. First, I attempt to clarify what (de-) or (re-)standardisation means. Second, I outline some theoretical approaches which support the de-standardisation hypothesis, and then argue for an alternative view from a theoretical perspective. Third, I present some empirical evidence. I present own findings for the German case and results from comparative research as found in the literature. Different trends can be identified, including cases of de-standardisation, of changing patterns without de-standardisation emerging, and of re-standardisation. Finally, I draw some conclusions. Having looked closely at what actually happens in European countries I propose that alternatives to the de-standardisation hypothesis must be considered. In particular, re-standardisation as a systematic phenomenon in demographic trends has to be considered more seriously than has so far been the case.

## 5.2 De-Standardisation and Re-Standardisation in the Life Course: Some Conceptual Clarifications

Before I discuss the de-standardisation issue from a theoretical and an empirical point of view I start with some conceptual clarifications.

### *5.2.1 The Concept of Standardisation and De-Standardisation*

When referring to de-standardisation of all or part of the life course, Brückner and Mayer (2005) mean that "life states, events, and their sequence can become experiences which either characterize an increasingly smaller part of a population or occur at more dispersed ages and with more dispersed durations" (Brückner and Mayer 2005, 32f). In Table 5.1, I seek to clarify in more detail what factors the level of standardisation of all or part of the life course, such as the family life

**Table 5.1** Dimensions of the level of standardisation of the life course

| | Dimension | Social determinants | Measured by |
|---|---|---|---|
| Events/ transitions | 1. Incidence of biographical events (state transitions) | Norms, institutions, and structural opportunities in regard to developmental goals and social roles | Coverage rates of biographical events |
| | 2. Age structure of biographical events (state transitions) | | Age specific state-transition rates; Variance of the age distribution of transitions |
| | 3. Interrelation between biographical events (state transitions) | Coupling of life course transitions and path dependency | Age span between transitions; Event dependency of transition rates |
| States | 4. (Age specific) diversity of biographical states | Norms, institutions, and structural opportunities in regard to social roles | Entropy of the distribution of biographical states |
| | 5. (Age specific) interrelation between biographical states | Coupling of states in different domains of the life course | Correlation between current biographical states |
| (Sub-) trajectories | 6. Reversibility of life events and stability and number of biographical states over time | Norms, institutions, and structural opportunities in regard to the whole structure of the life course | Number of repeated life events, Rates of exit and re-entry |
| | 7. Diversity of biographical (sub-) trajectories in terms of order and variety of biographical events | Coupling of sub-trajectories and path dependency | Entropy of the distribution of biographical state trajectories(order and variety of biographical events and states) |

trajectory or the status passage of transition to adulthood, depends upon. Several dimensions are distinguished (cf. Huinink and Konietzka 2003). I will not comment on them in detail.

In this context, de-standardisation means that the level of standardisation is declining in one of these seven dimensions. This can be, for example, increasing entropy[1] of age distributions, decreasing correlations between the occurrence of biographical states or events, or increasing entropy of the distribution of types of biographical trajectories.

[1] The entropy in this case can be conceived as measure for how close the distribution of a discrete variable (age in years, biographical state) is to the uniform distribution. If the variable is uniformly distributed the entropy is maximized—which means maximum de-standardisation. The entropy in case of a one-point distribution is 0.

The definition by Brückner and Mayer primarily refers to dimensions one to five, and partly to dimension six in Table 5.1. In the literature, other measures of the level of standardisation are proposed which mainly deal with trajectories or parts thereof (dimensions six and seven). Elzinga and Liefbroer (2007) analyse the turbulence and dissimilarity of trajectories, or the variation of pathways to adulthood. Elzinga and Liefbroer (2007, 232) "consider series [meaning trajectories, JH] that have many distinct states and many state changes as being more turbulent than series with fewer distinct states and/or fewer state changes". This concept is a measure of reversibility of life events and stability of biographical states in trajectories (dimension six in Table 5.1). The dissimilarity of trajectories and the variation of types of trajectories, which is based on a typology of typical patterns, are measures of diversity of trajectories (dimension seven in Table 5.1).

### *5.2.2 "Modes" of De-Standardisation and Re-Standardisation*

Conceptually, one can distinguish two modes of de-standardisation:

1. *The transitory case*
   De-standardisation can be just a transitory phenomenon caused by a special characteristic of the dynamics of pattern change (pattern-transition hypothesis). It is a passing phenomenon which occurs when the shift from one pattern to another does not encompass the whole population simultaneously but part by part like in the case of pattern diffusion. During the period of change an increasing part of the population experiences the new pattern while the other part still follows the old one. Consequently during the initial phase of pattern change the diversity increases (de-standardisation) but after a certain period of time it decreases again (re-standardisation) because finally the new pattern has been established in the whole population. If the pattern shifts gradually throughout the whole population simultaneously, change without any de-standardisation can take place. This, however, is an exception.
   De-standardisation of this transitory kind might also be caused by a transitory deviation from a pattern, disappearing after a certain period of time when in the end the old pattern survives. This can happen during and after historical events which have a major age specific impact on individual life courses, such as wars or economic crises.
2. *The irreversible case*
   De-standardisation might evolve as an ongoing trend of pattern dissolution over time (de-standardisation hypothesis). In this case, de-standardisation is not reversible in the long run. It means that a certain pattern is being diluted, without leading to a new one or a return to the previous one.
   The mode of transient de-standardisation makes clear that re-standardisation is a possibility. For example, one dominant age at a biographical event is, after a time,

simply "replaced" by a different one; or one dominant pathway to social independence from parents is "replaced" by another pathway. Re-standardisation means that one or more of the dimensions of standardisation in Table 5.1 is once again increasing: e.g., age patterns narrow, the connection between life events strengthens, a certain kind of new living arrangement in a certain age period is gaining dominance, or a sequence of events is appearing again. Corresponding to the definition of de-standardisation by Brückner und Mayer, re-standardisation means that life states, events, and their sequence become experiences which (again) either characterise an increasingly larger part of a population, or occur in more narrow age spans and with more uniform durations.

## 5.3 Theoretical Considerations

This section provides a brief overview of some theoretical arguments in the literature supporting the de-standardisation hypothesis. Some of these arguments have a particular focus on transition to adulthood. In the social sciences, the transition to adulthood is traditionally conceived of as a process in individuals' life courses which is connected with typical transition markers, such as starting and completing education, starting a job or career, leaving the parental home, forming a union, and a family. It is directed towards the goal of gaining individual economic and social independence by the young adult.

### *5.3.1 Sociological and Demographic Considerations*

The concept of the *Second Demographic Transition* (SDT) promotes the idea of a new period of demographic change affecting in particular the status passage of transition to adulthood. While the First Demographic Transition led to a highly standardised pattern of bourgeois family dynamics, labelled the "golden age of marriage", these patterns dissolve during the Second Demographic Transition. The main trends have already been mentioned: declining family size, increasing age at family formation, increasing relevance of non-marital living arrangements and extra-marital childbearing, and declining stability of couples and rising divorce rates. These trends, it has been argued, are the result of ongoing secularisation, shifts in value orientations, and an increasing relevance of higher level goals, like self-realisation and self-fulfilment in the lives of individuals (van de Kaa 1987; Lesthaeghe and Moors 2000; Lesthaeghe and Surkyn 2007). The question of de-standardisation has not been addressed explicitly in this context. In a recent article, Lesthaeghe and Surkyn made the following prediction regarding the SDT: "The normative and institutional props of traditional union formation and household structures will systematically weaken in all societies that move in the direction of egalitarian and democratic systems governed by the respect for individual choice. This implies that other forms of union formation will expand in the wake of such ideational developments" (Lesthaeghe and Surkyn 2007, 113). No explicit arguments have been made so far that the SDT could pave the way for new patterns to emerge, such as new standards of family formation and living arrangements.

That is why many scholars draw the conclusion that the SDT should, by and large, lead to more diversity. Corijn and Klijzing (2001) use assumptions about the SDT approach and the modernisation paradigm when assessing change in the transition to adulthood. In regard to age graduation, they generally propose that the age-relatedness of markers of transition to adulthood is weakening. However, they argue that age-relatedness of career related transitions remains stronger than age-relatedness of family related transitions. This is true because education and income capacity are the most important requirements for a successful life, and the institutional regulations and the normative expectations in regard to the time span for achieving these goals are quite restrictive. They also assume that de-standardisation should not surpass the age of thirty. The level of de-standardisation between different European countries should remain different. In regard to the interrelatedness of different dimensions of the transition to adulthood, they conventionally argue for an incompatibility between education and parenthood—more so for women than for men, and weakening with increasing age, by different degrees in European countries.

Billari and Wilson (2001) also deal with the question of diversity vs. similarity in the shift towards greater de-standardisation in the transition to adulthood in different European countries. They also do not support the hypothesis of convergence between European countries. By and large, they seem to take for granted that there is a general trend towards a de-standardisation of this status passage. In any case, other alternatives are not addressed when they conclude with the following: "Rather than seeing life course changes as being driven by strongly determined trends towards country-level convergence, of either the simple form or the individualization variety, perhaps we should see recent trends as being 'enabling'? With increasing wealth in European countries many of the material constraints on life course transitions have been reduced or removed. Thus, individuals in different societies are freer than ever before to give free rein to their preferences" (Billari and Wilson 2001, 14).

In sociology, a prominent position arguing for the change in individual life courses is based on the hypotheses of *de-institutionalisation* and ongoing *social differentiation* in late modern societies (Brückner and Mayer 2005). De-institutionalisation means "that states, stages, events, and transitions, which at earlier times were clearly differentiated [by institutional regulations; JH], are now being reintegrated or fused" (Brückner and Mayer 2005, 32). The change evolves after a period of institutionalisation and standardisation during modernisation (Kohli 2003). Currently, modern institutions like the bourgeois family system are eroding. Life courses are less institutionally predetermined. Young adults do not follow traditional social norms in planning their lives. Instead, they tend to optimize strategies in a far more flexible and autonomous way to make the most of a society that is now rich in opportunities and risks.

Social differentiation "refers to the process where the number of distinct states or stages across the life course increases" (Brückner and Mayer 2005, 33). It could be added that social differentiation also leads to a de-coupling of different dimensions of biographical states. The areas of activity and individual engagement they refer to increasingly follow their own procedural logic, with their own demands and options for individual actors.

Brückner and Mayer (2005, 35) note that processes of institutionalisation and de-institutionalisation could go hand in hand. This is likely to be particularly true in a highly functional differentiated society. While no general trend can be observed, they assume a movement towards de-standardisation and increasing flexibility in all fields of life .

Elzinga and Liefbroer (2007) start with the same concepts of differentiation and de-standardisation employed by Brückner and Mayer. They also assume that the mobility of young adults between different states has increased among recent cohorts. This hypothesis is tested by analysing family life trajectories in various European countries. The authors also propose that trajectories are less similar between young adults, and that no dominant type of family trajectory now exists. They also assume a considerable variation between different countries in Europe, anticipating that variations will be greatest in social democratic welfare states and smallest in the Mediterranean and East European countries.

Finally, an influential hypothesis in sociology proposes a new shift in *individualisation* in late-modern societies (Beck 1986; Shanahan 2000). Generally, individualisation means that individual actors are, in planning and mastering their life course, no longer tied to normative institutions, community related commitments, and traditional support systems. This implies a transformation from community type organised social contexts to contexts centred on the market and on formal regulation-based systems of co-ordination and integration. The "new individualisation" thesis assumes that women are liberated from the traditional roles of mother and housewife, and gain better access to the labour market. Furthermore, the modern institutions of the life course, like the bourgeois family pattern which resulted from individualisation processes during modernisation itself (Kohli 1985) lose their impact. Individual agency, flexibility, and choice gain relevance. The transition to adulthood can be expected to become more variable with the transition markers becoming decompressed (Shanahan 2000, 671), and the sequencing of markers becoming more diverse.

### *5.3.2 A Psychological View*

An approach coming from developmental psychology also argues in favour of more de-standardisation. Arnett introduces the term "emerging adulthood" (Arnett 2000, 469), a period defined as being "neither adolescence nor young adulthood".

Sociologists would say there is some anomic situation, without specific social roles and normative expectations to be obeyed. Emerging adults are becoming independent of their parents, and social control by parents has decreased considerably, yet these young people still have not taken on the "responsibilities that are normative in adulthood". The educational system has expanded, and marriage and childbirth have been delayed. Perhaps because they follow new age norms, economic pressure on young adults has been moderate. The result is "a high degree of demographic diversity and instability, reflecting the emphasis on change and exploration (in areas of love, work, and worldviews; JH)" (Arnett 2000, 471).

Arnett's arguments could well support the hypothesis that there is a de-standardisation of transition to adulthood, at least as far as the level of turbulence in trajectories during this period of the life course is concerned. Accepting that de-standardisation is occurring does not contradict the notion that, in terms of age and biographical states, re-standardisation and a rather standardised way into and out of this period could also be taking place. Emerging adulthood has itself become a period of the life course where the weak norms are normatively supported. The level of diversity is therefore a function of the supply of attractive opportunities to extend exploration; here we can draw links to the issues of differentiation and societal and technological development.

### 5.3.3 *Conclusions*

All approaches, as diverse as they are, generally appear to favour assumptions of an ongoing de-standardisation of the transition to adulthood. In addition to postponement of events, their age-relatedness should have declined, biographical disorder should have increased, and it should be possible to observe a decoupling of life events during the early adult life course (Furstenberg et al. 2005). Alternatives are not discussed. From my point of view, these concepts and theoretical approaches suffer from three deficits:

1. For the most part, they take for granted that de-institutionalisation and differentiation lead to the dissolution of life course regimes, without considering any concrete ideas of restructuring trends.
2. They do not differentiate between the various modes of de-standardisation, as I proposed.
3. They do not consider explicitly trends of re-standardisation, even though a hypothesis pertaining to this could easily be implemented in these theories.

In discussing the de-standardisation hypothesis, a general argument could be made that social change can be conceived as a sequence of periods of stability or standardisation, and periods of more or less rapid change accompanied by (transient) de-standardisation. In addition, it could be argued that scientific, economic, and social developments lead to an increase in opportunities and abilities, which support an *underlying* trend of increasing diversity in human behaviour and individual life courses. Both aspects must be considered. Given the general nature of this discussion, I will not address these issues in great detail, but will instead restrict myself to some simple arguments.

De-standardisation can be conceived of as a possible structural or demographic consequence of de-institutionalisation and differentiation (Huinink and Konietzka 2003). But the question of de-standardisation is more complicated than is frequently assumed (Huinink and Wagner 1998; Brüderl 2004). De-institutionalisation may be a necessary condition for de-standardisation, but it is not sufficient. If we assume that people act (quite) rationally in a given set of living conditions, we can propose the following: Even without social norms governing social action, it should be possible to find common behavioural patterns reflecting the optimal way to deal with circumstances of action and life course planning. Therefore, de-standardisation can mean that the circumstances in which young men and women plan their lives have become increasingly diverse. At the same time, however, de-institutionalisation itself might just be a transient phenomenon occurring in a phase of changing transition patterns. De-standardisation could diminish as soon as new institutional patterns take over after a period of re-institutionalisation (pattern-transition or re-standardisation hypothesis).

In exploring the first point, it is interesting to look at the theoretical concept Blossfeld and his research group proposed in studying the consequences of globalisation on individual life courses. They argue that the restructuring of local economies and demographic change in many countries put welfare state institutions and labour markets under pressure. On the individual level, biographical uncertainties arise affecting individual life plans and instabilities. New risks in educational and occupational careers emerge as a new threat to a promising transition to adulthood (Blossfeld et al. 2005). Following an action theoretic model of bounded rationality, they state that young men and women cope with this uncertainty under the given institutional conditions. One central hypothesis is, then, "that the uncertainty generated by globalisation at the socio-structural level reduces or delays the propensity of youth to enter long-term binding commitments such as partnership and parenthood" (Mills and Blossfeld 2005, 16). However, they do not propose an increasing de-standardisation of early life courses in general, but rather suggest that there is considerable diversity caused by different institutional conditions in the mastering of increasing uncertainty.

In conclusion, we can state that, in some life domains or dimensions of transition to adulthood, irreversible trends of de-standardisation may emerge, particularly in the case of a stable interplay between de-institutionalisation and social differentiation. Sometimes de-standardisation does not occur. In other cases, de-standardisation is transient, and comes to an end when new patterns are established that are grounded on new institutions or new structural typicality in the lives of young individuals. The specific circumstances surrounding the achievement of the different goals on the way to adulthood determine what can theoretically be expected. Which version might be most probable for which dimension of transition to adulthood? This question is difficult to answer, and I cannot present an elaborated theory at this point. Instead, let us look at empirical examples that may supply us with evidence supporting one of these cases.

## 5.4 Some Empirical Evidence

My empirical analyses focus on four of the dimensions previously mentioned in Table 5.1. The age structure of biographical events, the interrelation between biographical events, the sequence of biographical events, and the diversity of whole trajectories. I continue to focus on the status passage of transition to adulthood.

### *5.4.1 The Age Structure of Biographical Events*

Let us start with a comparison of the age at biographical transitions leading to social independence from parents between different cohorts. Beginning with the West German case, I look at first household formation, first consensual union, and family formation.

In Table 5.2, parameters of the age distribution at first household formation for men and women of seven West German cohorts are displayed. Men's median age at first household formation decreased until the cohort 1959–1961, from age 26 to 23. In later cohorts, median ages rose again, though only by less than 1 year. For women, we observe a decline of the median age from 24 to 20.5 years between cohorts 1929–1931 and 1954–1956. Again, in the younger cohorts, the median age increased by about 1 year. Over the same period, the "time corridor" of leaving home narrowed over the cohorts. After a considerable decline, a slight increase in

**Table 5.2** Age at first household formation by cohort and gender in West Germany

| | Cohort | | | | | | |
|---|---|---|---|---|---|---|---|
| | 1929–1931 | 1939–1941 | 1949–1951 | 1954–1955 | 1959-1961 | 1964 | 1971 |
| *Men* | | | | | | | |
| 1. Quartile | 23.25 | 22.50 | 21.50 | 21.33 | 21.08 | 21.08 | 21.42 |
| 2. Median | 26.00 | 25.25 | 24.08 | 23.58 | 23.00 | 23.75 | 24.00 |
| 3. Quartile | 30.50 | 29.00 | 27.33 | 26.50 | 26.17 | 26.33 | 27.00 |
| *Women* | | | | | | | |
| 1. Quartile | 21.33 | 20.50 | 19.42 | 19.08 | 19.08 | 19.58 | 19.83 |
| 2. Median | 24.33 | 22.50 | 21.08 | 20.50 | 20.67 | 21.33 | 21.67 |
| 3. Quartile | 29.42 | 25.25 | 23.33 | 22.67 | 23.00 | 23.67 | 24.33 |

*Source* German Life History Study, Kaplan–Meier Estimates [Huinink and Konietzka (2006), the changing impact of union formation on leaving home in Germany. A cohort analysis of interdependent life events in the transition to adulthood "unpublished"]

the age difference between the first and the third quartile can be observed for men in the 1971 cohort. Still, the age difference was only about 5.5 years for the youngest cohort in our sample, while it was more than 7 years in the oldest cohort. For women, the decrease in the age difference between the first and the third quartiles was even greater, falling from about 8 years in the cohort 1929–1931, to as low as around 3.5 years in the cohort 1954–1956, and then rising again to 4.5 years in the 1971 cohort. These figures suggest that there is little indication for de-standardised patterns of first household formation in terms of cohort-specific age variations in West Germany.

The same holds true if we look at the age at starting the co-residential union. In Table 5.3, results from Kaplan–Meier estimates of this age are displayed. I used data from the German Family Survey 2000. (Age is calculated as the difference between the year of the beginning of the partnership and the birth year). As can be seen in Table 5.3, there is little change in regard to the age structure of this event for men and women. No trend of age dispersion is apparent.

Finally, let us turn to the transition to first marriage and to the first child. I focus only on West German men and women. Again, I used data from the German Family Survey 2000. The respective figures are displayed in Table 5.4.

Both age at first marriage and age at first birth increased considerably over cohorts. Looking at the first marriage of West German women, we see that the age range widens over cohorts when the age at first marriage rises. The interquartile differences for women increase from 5.5 years in cohort 1944–1949 to 9.3 years in cohort 1962–1967. For men, we cannot compare interquartile differences between cohorts with the data used.

A different picture is apparent for age at first birth. Women of the cohort 1944–1949 show quite a standardised pattern of family formation at an early age. By contrast, age at first birth is quite dispersed among women of the cohort 1950–1955. It is a typical transitional cohort at a time when the trend towards delay of family formation was underway. In the younger cohort of 1962–1967, a new age graduation appears to have begun, though the delaying process was not completed. Late family formation in the late twenties was becoming increasingly

**Table 5.3** Age at first consensual union by cohort and gender in West Germany

| | Cohort | | | |
|---|---|---|---|---|
| | 1944–1949 | 1950–1955 | 1956–1961 | 1962–1967 |
| *Men* | | | | |
| 1. Quartile | 19 | 19 | 18 | 18 |
| 2. Median | 22 | 21 | 21 | 21 |
| 3. Quartile | 26 | 28 | 26 | 26 |
| *Women* | | | | |
| 1. Quartile | 17 | 17 | 17 | 17 |
| 2. Median | 19 | 19 | 19 | 19 |
| 3. Quartile | 21 | 22 | 22 | 22 |

*Source* German Family Survey (2000), Kaplan–Meier Estimates

**Table 5.4** Age at first marriage and first birth of West German men and women, by cohort

| | First marriage | | | | First birth | | | |
|---|---|---|---|---|---|---|---|---|
| | Cohort | | | | Cohort | | | |
| | 1944–1949 | 1950–1955 | 1956–1961 | 1962–1967 | 1944–1949 | 1950–1955 | 1956–1961 | 1962–1967 |
| *Men* | | | | | | | | |
| 1. Quartile | 22.5 | 23.1 | 23.9 | 25.2 | 23.8 | 26.3 | 26.1 | 26.3 |
| 2. Median | 25.8 | 27.5 | 28.3 | 30.5 | 28.8 | 32.2 | 31.6 | 32.6 |
| 3. Quartile | 32.7 | –[a] | –[a] | –[a] | –[a] | –[a] | –[a] | –[a] |
| *Women* | | | | | | | | |
| 1. Quartile | 19.5 | 20.2 | 20.7 | 22.0 | 20.4 | 21.5 | 22.3 | 23.5 |
| 2. Median | 21.5 | 22.6 | 23.8 | 25.5 | 23.4 | 26.0 | 26.1 | 27.4 |
| 3. Quartile | 25.0 | 28.7 | 29.2 | 31.3 | 30.5 | 33.1 | 31.3 | 32.1 |

[a] Not estimated

*Source* German Family Survey (2000), Kaplan–Meier Estimates

normal. The interquartile difference, which can be used as measure of the level of standardisation, changes over cohorts. For the cohort 1944–1949, it is 10.1 years, while for the cohort 1950–1955, the estimate is 11.6 years. For the cohort 1956–1961, the estimate is 9.0 years, and for the cohort 1962–1967, it is even smaller (8.6 years). This shows the transitory character of the emergence of an age-dispersed pattern of family formation in the cohort 1950–1955. Interestingly, we see that, for women, the findings support the idea that de-standardisation is just a transient phenomenon caused by the age shift for family formation, but not for marriage.

Billari and Wilson (2001) did similar calculations for different European countries, allowing us to make international comparisons. They estimated interquartile differences in ages at leaving home, first union, first marriage, and first birth. I only show part of one of their tables dealing with first marriage and first birth. We see in Table 5.5 that the patterns for West Germany match results from some other countries. In nearly all countries, the age at marriage is constantly de-standardising, particularly in Sweden. In the case of age at first birth, we often observe a nonlinear trend, as could be expected given the pattern change assumption.

### *5.4.2 Interrelation Between Biographical Events*

Now I turn to some examples of the interrelation between the markers of transition to adulthood and the sequencing of events. Currently in West Germany, relatively few young men or women first move in together at or after marriage [Huinink and Konietzka (2006), the changing impact of union formation on leaving home in Germany. A cohort analysis of interdependent life events in the transition to adulthood "unpublished"; Kley and Huinink (2006)]. As Fig. 5.1 shows, cohabitation has nearly completely taken over the role of marriage in this regard, more so for women than for men. But moving in with a partner still is the most frequent reason for leaving the parental home in West Germany.

It is possible to add figures on the dependency of different transition rates on leaving the parental household which nicely support these descriptive findings [Huinink and Konietzka (2006), the changing impact of union formation on leaving home in Germany. A cohort analysis of interdependent life events in the transition to adulthood "unpublished"]. I interpret this trend to be a kind of re-standardisation and re-institutionalisation following the de-institutionalisation of marriage. As has been shown in many analyses, a decoupling of first marriage and first birth can be observed in many European countries, as is the case for Germany. In Southern European countries like Italy, marriage still seems to be a prerequisite for leaving home. That might be one major reason for the unprecedented rise in the median age at leaving home there. Paradoxically, the stability of institutions is causing a change in patterns of transition to adulthood, leading to long delays in family formation.

**Table 5.5** Interquartile differences in ages at events in the transition to adulthood (years)

| Gender | Men | | | | Women | | | |
|---|---|---|---|---|---|---|---|---|
| Cohort | 1946–1950 | 1951–1955 | 1956–1960 | 1961–1965 | 1946–1950 | 1951–1955 | 1956–1960 | 1961–1965 |
| *First Marriage* | | | | | | | | |
| Austria | 7.2 | 6.6 | 7.5 | >11.3 | 4.5 | 5.5 | 8.1 | 7.4 |
| Belgium (Flemish) | – | 4.3 | 5.7 | >8.5 | – | 3.3 | 4.6 | 5.5 |
| Finland | 8.7 | >16.0 | >12.7 | >3.0 | 6.3 | 7.7 | 10.6 | >6.2 |
| France | 5.4 | 8.0 | >15.5 | >8.6 | 5.0 | 5.3 | 10.3 | >11.3 |
| Germany (West) | – | 10.4 | >12.2 | >7.0 | – | 7.2 | 8.8 | >9.7 |
| Italy | 5.4 | 5.6 | 8.5 | >8.3 | 4.7 | 5.4 | 7.8 | 8.3 |
| Portugal | 3.5 | 4.6 | 6.1 | 8.3 | 5.3 | 4.2 | 5.8 | 6.5 |
| Spain | 5.7 | 6.3 | 6.6 | >9.0 | 4.3 | 4.5 | 5.3 | 6.8 |
| Sweden | 15.8 | – | >6.1 | >1.3 | 10.6 | 15.2 | >6.1 | >3.6 |
| *First Birth* | | | | | | | | |
| Austria | 11.3 | 7.0 | 12.1 | >11.3 | 6.4 | 7.0 | 8.1 | 7.3 |
| Belgium (Flemish) | – | 8.0 | 7.6 | >5.6 | – | 5.4 | 6.3 | 6.5 |
| Finland | 10.7 | 11.4 | >11.7 | >2.9 | 8.2 | 7.7 | 9.5 | >5.3 |
| France | 8.1 | 10.7 | >8.5 | >7.3 | 7.0 | 6.6 | 8.5 | >10.8 |
| Germany (West) | – | >15.9 | >10.9 | >6.5 | – | 9.8 | 8.9 | >8.7 |
| Italy | 8.9 | 7.8 | 11.7 | >6.8 | 5.9 | 6.5 | 10.5 | 9.1 |
| Portugal | 4.9 | 5.8 | 7.5 | 8.9 | 5.7 | 4.7 | 6.1 | 6.8 |
| Spain | 6.2 | 7.7 | 7.7 | >6.7 | 4.9 | 5.2 | 7.0 | 7.9 |
| Sweden | 8.9 | – | >8.7 | >3.2 | 7.8 | 8.5 | 8.3 | >6.1 |

*Source* Family and Fertility Survey (Billari and Wilson 2001, 22)

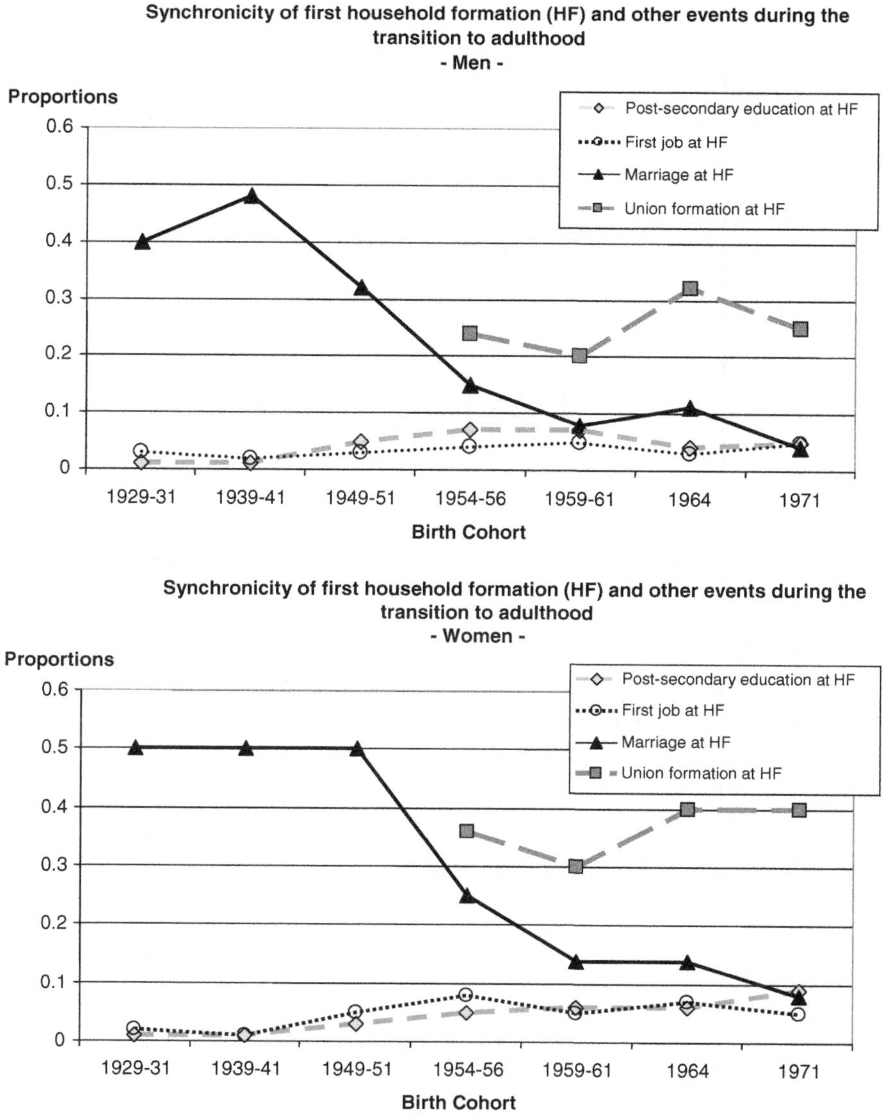


**Fig. 5.1** First household formation and other markers of transition to adulthood. *Source* German Life History Study [Huinink and Konietzka (2006), the changing impact of union formation on leaving home in Germany. A cohort analysis of interdependent life events in the transition to adulthood "unpublished"]

Our comparative outlook now deals with the sequencing of leaving home, of first union (cohabitation), and of first marriage. Again, we display the results of Billari and Wilson (2001) in Table 5.6. In countries which are assumed to be prototypes of de-standardisation, we find highly standardised patterns. The

**Table 5.6** Sequencing between events during transition to adulthood (years)

| Gender | Men | | | | Women | | | |
|---|---|---|---|---|---|---|---|---|
| Cohort | 1946–1950 | 1951–1955 | 1956–1960 | 1961–1965 | 1946–1950 | 1951–1955 | 1956–1960 | 1961–1965 |
| *Leaving home before first union (%)* | | | | | | | | |
| Austria | 47 | 49 | 48 | 55 | 40 | 41 | 47 | 38 |
| Belgium (Flemish) | – | 15 | 18 | 22 | – | 12 | 15 | 17 |
| Finland | 57 | 63 | 61 | 59 | 57 | 58 | 55 | 54 |
| France | 43 | 47 | 53 | 49 | 37 | 35 | 40 | 42 |
| Germany (West) | – | 54 | 51 | 53 | – | 37 | 42 | 44 |
| Italy | 31 | 21 | 27 | 32 | 13 | 12 | 16 | 14 |
| Portugal[a] | 12 | 13 | 20 | 19 | 14 | 18 | 18 | 20 |
| Spain | 9 | 12 | 19 | 28 | 4 | 6 | 8 | 14 |
| Sweden | 69 | – | 69 | 73 | 63 | 64 | 63 | 64 |
| *First union before first marriage (%)* | | | | | | | | |
| Austria | 46 | 52 | 83 | 82 | 31 | 45 | 62 | 71 |
| Belgium (Flemish) | – | 14 | 19 | 31 | – | 12 | 15 | 24 |
| Finland | 45 | 86 | 85 | 93 | 39 | 66 | 83 | 88 |
| France | 35 | 49 | 65 | 81 | 25 | 34 | 53 | 69 |
| Germany (West) | – | 49 | 67 | 74 | – | 40 | 58 | 68 |
| Italy | 4 | 11 | 14 | 15 | 4 | 6 | 8 | 11 |
| Spain | 9 | 12 | 19 | 28 | 4 | 6 | 8 | 14 |
| Sweden | 87 | – | 94 | 94 | 81 | 90 | 92 | 92 |

[a] Leaving home before first marriage

*Source* Family and Fertility Survey (Billari and Wilson 2001, 22)

estimations of Billari and Wilson show that Sweden is ahead in a re-standardisation in this regard. Other countries are on the way. In Sweden, the traditional patterns have diminished, resulting in new regularities. Formerly unconventional sequences of events now are obligatory. This is in accordance with the pattern–transition hypothesis. We do not know, however, whether this is caused by new social norms or by other factors.

### *5.4.3 Diversity of Biographical (Sub-)Trajectories*

#### Living Arrangements and Family Career

Relationships are arranged differently today than they were in the past. A pullback from marriage can be observed, and the age for starting a consensual union with a common household is rising, too. Living apart together and cohabitation are gaining relevance, as is living alone for a period of time. However, this does not mean that "anything goes". Cohabitation has evolved into a highly institutionalised step or long-term alternative to marriage.

But from the longitudinal view, a moderate shift in the diversity of the trajectory of living arrangements is shown for West Germany (Brüderl and Klein 2003; Brüderl 2004). Data from the German Family Survey 2000 show that the entropy of the distribution of living arrangements among young adults after age 22 increased over time, more so in large cities than in rural areas. One reason for this trend is the increasing proportion of men and women living without a partner at this age segment. This is a transient state in most cases, but obviously it occurs by a higher rate. Brüderl and Klein also show an increasing divergence from the traditional pattern in the pathways to adulthood. This can, but by no means must be seen as evidence of de-standardisation, because new patterns of standardisation could have emerged.

To provide an international comparison, I briefly come back to Elzinga and Liefbroer (2007), who studied family trajectories of young adults in different European countries using data of the Family and Fertility Survey. They compared cohorts 1945–1949, 1950–1954, 1955–1959, and 1960–1964. As mentioned previously, they developed new measures of turbulence, the similarity of trajectories and their variation over seven 'ideal types' of pathways.

They found that turbulence was rather stable over cohorts in most countries, but it rose in some places, and it even declined in two countries (Poland and Italy). The similarity between trajectories declined in nearly all countries, though only slightly in some of them. The level of variation between types of trajectories was also found to be rising in most countries. In Sweden, it has been declining over cohorts. Elzinga and Liefbroer argue that, in this case, a new standard might be emerging, and they refer to re-standardisation. They conclude: "These results offer strong support to the idea that the family life trajectories of young adult women all across the Western world are becoming more de-standardized. […] the decrease in variation in family life trajectories observed in Sweden and, to a lesser extent, also in the Netherlands, suggests that new standards may be emerging in these countries. In that sense, it is not impossible that a process of re-standardization will occur in countries that have experienced the Second Demographic Transition" (Elzinga and Liefbroer 2007, 246–247).

Do these results contradict the findings of Billari and Wilson? I do not think so, assuming that turbulence is not equated with de-standardisation. Taken together, the results of these two studies confirm the assumption that the stability of trajectories during the transition to adulthood has been declining. Yet, we nevertheless observe standardised parts of sequences and transitions, mainly based on new standards of ageing and sequencing.

## Apprenticeship and Labour Market Entry

Finally, we take a brief look at the transition to economic autonomy in Germany: starting and completing post-school training or apprenticeship, and starting a job or career. (In Germany, vocational education of this kind is very important: it combines apprenticeship in a company with part-time education in special vocational schools that provide the theoretical education to the company-based practical training). The de-standardisation hypothesis proposed by Arnett could prove useful in explaining these trajectories, because his hypothesis is more strongly related to processes of differentiation, and is motivated by his concept of emerging adulthood.

A certain trend towards de-standardisation can, in fact, be shown in regard to vocational training and apprenticeships. The findings of Konietzka (1999) for the West German case, which are based on data from the German Life History Study, show that the percentage of young men and women who started more than one apprenticeship increased significantly from the cohorts of 1929–1931 to the cohorts of 1959–1961. However, the percentage of men who completed more than one apprenticeship was stable over cohorts. It rose only slightly for women. This could mean that, in younger cohorts, a larger share of young men and women try out different kinds of apprenticeship before completing one that best fits their interests (cf. Arnett 2000). The increase can be a function of improving opportunities as well as higher uncertainties about their life plans. While the former would support a trend to irreversible de-standardisation, this is not true for the latter.

Konietzka is also able to show that the transition to the first job, even when it occurs in times of economic crisis, is still quite regular among West German cohorts, and is only gradually becoming fuzzy. No de-standardisation is apparent. But West Germany might be a special case when compared with other European countries, for which I cannot present similar studies.

## 5.5 Conclusions

The brief overview of studies of the changing transition to social and economic autonomy of adolescents and young adults shed light on both the processes of transient de-standardisation as a consequence of pattern change, and the trends towards more diversity, flexibility, and variety of trajectories during the early life course. It is true that the latter is dominant in Europe. In line with Arnett's theory, there are hints of a more pronounced period of experimentation. We also know that phases of living alone gain relevance. The level of the institutionalisation of intimate relationships has changed, with the strength of commitment in the early stages of relationships being lower than in the past. The diversity, at least in the longitudinal perspective (trajectories), has become larger, and yet more differentiation of living arrangements can be assumed if the numerous versions of arrangements where partners are regionally mobile are taken into consideration. It should also be noted that the prevalence of highly committed living arrangements has also declined, representing yet another indication of de-standardisation.

However, this does not mean that there are no signs of new institutionalisation and re-standardisation. For example, cohabitation has virtually become the standard step into a co-residential household in a growing number of countries. Even family formation out of wedlock is increasingly becoming a standard behaviour in some countries like Sweden and East Germany. The age at which certain markers of transition to adulthood take place has changed, but in many cases not dispersed. The age structure of events like marriage and family formation has changed, but also re-stabilised.

De-standardisation might have its limits. For example, partnership, however it is realised, is still an important state to be reached quite early in adult life. New schedules for young adults might emerge. Therefore, enduring de-standardisation is not the only alternative for the future. For instance, that social norms regarding leaving home still matter, but only in a very restricted way, is nicely shown by Billari and Liefbroer (2007).

This is why empirical alternatives to the de-standardisation hypothesis must be sought. In particular, re-standardisation as a systematic phenomenon in demographic trends has to be considered more seriously than has so far been the case. Theoretically, it is necessary to supplement the presented approaches with a profound concept of social change which corresponds to general assumptions about the logic of processes of pattern transitions.

## References

Arnett, J.-J. (2000). Emerging adulthood. A theory of development from the late teens through the twenties. *American Psychologist, 55*(5), 469–480.

Beck, U. (1986). *Die Risikogesellschaft*. Frankfurt/M.: Suhrkamp.

Billari, F. C. (2004). Becoming an adult in Europe: A macro(/micro)-demographic perspective. *Demographic Research, 3*(2), 15–44. Special Collection.

Billari, F. C., & Liefbroer, A. (2007). Should I stay or should I go? The impact of age norms on leaving home. *Demography, 44*(1), 181–198.

Billari, F. C., & Wilson, C. (2001). *Convergence towards diversity? Cohort dynamics in the transition to adulthood in contemporary Western Europe*, MPIDR WP 2001–039. Rostock: Max Planck Institute for Demographic Research.

Blossfeld, H.-P., Klijzing, E., Mills, M., & Kurz, K. (Eds.). (2005). *Globalization, uncertainty and youth in society*. Oxford: Routledge.

Brückner, H., & Mayer, K. U. (2005). De-standardization of the life course: What might it mean? And if it means anything, whether it actually took place. In R. Macmillan (Ed.), *The structure of the life course: Standardized? Individualized? Differentiated? (Advances in Life Course Research, Vol. 9)* (pp. 27–53). Amsterdam: Elsevier.

Brüderl, J. (2004). Die Pluralisierung partnerschaftlicher Lebensformen in Westdeutschland und Europa. *Aus Politik und Zeitgeschichte, 19*, 3–10.

Brüderl, J., & Klein, T. (2003). Die Pluralisierung partnerschaftlicher Lebensformen im Kohortenvergleich. In J. Marbach & W. Bien (Eds.), *Partnerschaft und Familiengründung* (pp. 189–217). Opladen: Leske+Budrich.

Corijn, M., & Klijzing, E. (2001). Transitions to adulthood in Europe: Conclusions and discussion. In M. Corijn & E. Klijzing (Eds.), *Transitions to adulthood in Europe* (pp. 313–340). Dordrecht: Kluwer Academic Publishers.

Elzinga, C. H., & Liefbroer, A. C. (2007). De-standardization of family life trajectories of young adults: a cross-national comparison using sequence analysis. *European Journal of Population, 23*(3–4), 225–250.

Furstenberg, F. F., Rumbaut R. G., & Settersten, R. A. (2005). On the frontier of adulthood: Emerging themes and new directions. In R. A. Settersten, F. F. Furstenberg, & R. G. Rumbaut (Eds.). *On the frontier of adulthood: Theory, research, and public policy* (pp. 3–28). Chicago: University of Chicago Press.

Huinink, J., & Konietzka, D. (2003). Die De-Standardisierung einer Statuspassage? Zum Wandel des Auszugs aus dem Elternhaus und des Übergangs in das Erwachsenenalter in Westdeutschland. *Soziale Welt, 54*(3), 285–312.

Huinink, J., & Wagner, M. (1998). Individualisierung und Pluralisierung von Familienformen. In F. Jürgen (Ed.), *Die Individualisierungsthese* (pp. 85–106). Opladen: Leske+Budrich.

Kley, S., & Huinink, J. (2006). Die Gründung des eigenen Haushalts bei Ost-und Westdeutschen nach der Wiedervereinigung. *Zeitschrift für Bevölkerungswissenschaft, 31*(1), 127–154.

Kohli, M. (1985). Die Institutionalisierung des Lebenslaufs. Historische Befunde und theoretische Argumente. *Kölner Zeitschrift für Soziologie und Soziapsychologie, 37*(1), 1–29.

Kohli, M. (2003). Der institutionalisierte Lebenslauf: ein Blick zurück und nach vorn. In J. Allmendinger (Ed.). *Entstaatlichung und soziale Sicherheit. Verhandlungen des* 31. *Kongresses der Deutschen Gesellschaft für Soziologie in Leipzig* 2002 (pp. 525–545). Opladen: Leske und Budrich.

Konietzka, D. (1999). *Ausbildung und Beruf. Die Geburtsjahrgänge* 1919-1961 *auf dem Weg von der Schule in das Erwerbsleben.* Opladen: Westdeutscher Verlag.

Lesthaeghe, R., & Moors, G. (2000). Recent trends in fertility and household formation in the industrialized world. *Review of Population and Social Policy, 9*, 121–170.

Lesthaeghe, R., & Surkyn, J. (2007). When history moves on: The foundations and diffusion of a second demographic transition. In R. Jayakody, A. Thornton, & W. Axinn (Eds.), *International family change, ideational perspectives* (pp. 81–117). New York: Lawrence Erlbaum.

Mills, M., & Blossfeld, H.-P. (2005). Globalization, uncertainty and the early life course. A theoretical framework. In H.-P. Blossfeld, E. Klijzing, M. Mills, & K. Kurz (Eds.), *Globalization, uncertainty and youth in society* (pp. 1–24). New York: Routledge.

Shanahan, M. J. (2000). Pathways to adulthood in changing societies: Variability and mechanisms in life course perspective. *Annual Review of Sociology, 26*, 667–692.

van de Kaa, D. J. (1987). Europe's second demographic transition. *Population Bulletin, 42*(1), 1–59.

# Chapter 6
# Europe, the Oldest-Old Continent

**Roland Rau, Magdalena M. Muszyńska and James W. Vaupel**

## 6.1 Introduction

Europe is growing older. For 2010, the United Nations calculate the old-age dependency ratio in Europe to have reached the value of 24 (see Table 6.1).[1] This implies that there are about four people at working ages (15–64) for each person at retirement age and above (65+). According to these UN estimates, Europe is the forerunner in aging among the world's regions. The world as a whole will reach this level of old-age dependency in the middle of the twenty-first century. By this time, the proportion of elderly in relation to people aged 15–64 would have increased further in Europe to reach the level of two people at working ages for each "pensioner".

These population dynamics are shaped—as usual in demography—by three sources of population flows: fertility, migration, and mortality. In the past, fertility played the primary role in changing the age structure of populations. For instance, in the late 1960s, the United Nations (1968: 7) noted that "As long as mortality remains within the 'universe' defined above, variations in mortality have little

[1] The old-age dependency ratio is the percentage of population aged 65 years and above to the population aged 15–64.

R. Rau (✉)
Institute for Sociology and Demography, University of Rostock, Rostock, Germany
e-mail: roland.rau@uni-rostock.de

M. M. Muszyńska
Institute of Statistics and Demography, Warsaw School of Economics, Warsaw, Poland
e-mail: mmuszyns@sgh.waw.pl

J. W. Vaupel
Laboratory of Survival and Longevity and the Laboratory of Evolutionary Biodemography,
Max Planck Institute for Demographic Research, Rostock, Germany
e-mail: jwv@demogr.mpg.de

G. Neyer et al. (eds.), *The Demography of Europe*,
DOI: 10.1007/978-90-481-8978-6_6,

**Table 6.1** Old age dependency ratio by world region in selected years

| Region | Years | | | | | |
|---|---|---|---|---|---|---|
| | 1950 | 1975 | 2000 | 2010 | 2025 | 2050 |
| Africa | 6 | 6 | 6 | 6 | 7 | 11 |
| Asia | 7 | 7 | 9 | 10 | 15 | 27 |
| Europe | 13 | 18 | 22 | 24 | 32 | 47 |
| Latin America & the Caribbean | 6 | 8 | 9 | 11 | 16 | 31 |
| Northern America | 13 | 16 | 19 | 19 | 29 | 36 |
| Oceania | 12 | 12 | 15 | 17 | 23 | 30 |
| World | 8 | 10 | 11 | 12 | 16 | 25 |

*Source* Medium Variant Projection of United Nations (2009)
The old age dependency ratio is 100 times the population 65+ divided by the population 15–64

effect on the age structure of a population, the latter being determined primarily by variations in fertility".[2] In recent decades, however, improved chances of survival are the primary reason for changes in age structures of populations in highly developed countries. As Preston et al. (1989: 691) showed for the United States and Sweden: "the dominant factor in current aging in these countries is a history of declining mortality".[3]

Thus, in this chapter, we discuss mortality developments and the various faces of demographic aging, illustrating them by examples from Norway, Denmark, Sweden and the former East Germany. We chose those countries mainly for the reason that this chapter has been written in honor of Jan Hoem—who lived for extensive periods of time in each of these countries. Furthermore, mortality patterns in these countries could be considered representative for Europe in general, at the same time as they show some interesting variation between countries. For example, Norway and Sweden have been among the countries with the highest life expectancy in the world for extended periods of time. In contrast, Danish women diverted from the positive trends in the early 1980s after which survival improvements remained marginal for about two decades. The opposite development has been experienced by the last country of analysis: after longevity stagnation since the 1970s, the former GDR experienced unprecedented gains in life expectancy after German reunification and the socio-economic transformation of the 1990s.

The main part of this contribution presents developments in survival and longevity among our four selected countries. Furthermore, we show that people not only live longer in general but also live longer in good health as the number of

[2] The "universe" in mortality variations refers to: "Starting from a level of mortality corresponding to an expectation of life at birth of about 25 years, the populations of the various areas have been moving with varying degrees of rapidity towards a level corresponding to an expectation of life at birth of about 75 years" (United Nations 1968: 7).

[3] According to their estimations, the average annual increase in the mean age of the female population in Sweden between 1980 and 1985 was mainly attributed to mortality change (≈61 %), while migration (≈35 %) and fertility (≈4.4 %) played only minor roles.

healthy years of life is growing along with increasing longevity. In our summary we discuss whether our results can be considered representative for other Western countries.

We end this chapter with a research outlook. First, we describe how life expectancy could develop if recent trends persist. Secondly, we stress the fact that the positive developments in survival and longevity also put demands on society as a whole, e.g. regarding financing of social security and old-age pensions. We propose that the paradigm of a strict distinction between the periods of working and retirement life may need to be abandoned in order to alleviate the burden of population aging for the welfare state.

## 6.2 Numbers of Old People

The most apparent sign of an aging population is the increasing number of people at advanced ages. In Table 6.2 we present the number of people aged 65–79, 80–99, centenarians (100+), and the sum of people 65+ for our four and selected other developed countries in 1939, 1960, and 2005. Between 1939 (Jan Hoem's birth year) and the last year of available data for all countries (2005), we also consider 1960, because it is approximately the year of onset of mortality decline among the oldest-old in many countries (see, for instance, Kannisto 1994; Vaupel 1997).

In 1939, when Jan Hoem was born, about 250,000 people aged 65 or older—slightly larger than the population of Rostock—lived in the whole country of Norway. In 2005, there were about 678,000 Norwegians in that age range. Similarly, the number of old people increased by almost a factor of three in Denmark and Sweden, the other two Nordic countries of analysis. This translates into an increase in the proportion of elderly in Norway from 8.6 to 14.7 %. As shown in Table 6.2, Norway is still rather "young" as compared to the other selected countries. In 2005, for example, almost every fifth person was 65 years or older in Eastern Germany. Together with Italy and Japan, the former GDR belongs to the countries with the highest proportion of elderly people in the world.

Even more remarkable is the increase in the number of people aged 80 or older (often labeled the "oldest-old"), including centenarians. In 1939, about one percent of the population in the selected countries had celebrated their 80th birthday. In most countries presented, 66 years later the "oldest-old" (80+) made up about 4–5 % of the total population. Similarly, just 70 years after 1939, meeting a centenarian has become considerably more common place: while there was about one centenarian in every 100,000 people in 1939, nowadays one can expect to find a centenarian for every 10,000 people.

By plotting population pyramids we illustrate in Fig. 6.1 the changing age structure, by single ages for women and men in Sweden in 1939, 1960, and 2005, respectively. (For the other countries under study the population pyramids have similar shapes.) Additionally, we show the age distribution in 2050 according to a projection by Statistics Sweden. Horizontal reference lines have been added at

**Table 6.2** Absolute and relative number of old people, both sexes combined, in selected countries in 1939, 1960, and 2005

Absolute frequencies

| Country | Ages 65–79 | | | Ages 80–99 | | | Ages 100+ | | | Ages 65+ | | |
|---|---|---|---|---|---|---|---|---|---|---|---|---|
| | 1939 | 1960 | 2005 | 1939 | 1960 | 2005 | 1939 | 1960 | 2005 | 1939 | 1960 | 2005 |
| Canada | 615,661 | 1,135,154 | 3,085,113 | 105,042 | 222,421 | 1,093,953 | 138 | 270 | 4,098 | 720,841 | 1,357,845 | 4,183,164 |
| Denmark | 250,200 | 406,341 | 591,632 | 40,502 | 73,390 | 220,243 | 10 | 19 | 628 | 290,712 | 479,750 | 812,503 |
| France | 3,529,500 | 4,385,139 | 7,231,222 | 531,429 | 908,010 | 2,744,660 | 101 | 368 | 11,589 | 4,061,030 | 5,293,517 | 9,987,471 |
| Germany | – | – | 11,810,469 | – | – | 3,459,869 | – | – | 8,253 | – | – | 15,278,591 |
| Germany-East | – | 2,025,012 | 2,576,659 | – | 314,822 | 688,734 | – | 31 | 1,758 | – | 2,339,865 | 3,267,151 |
| Germany-West | – | 5,153,914 | 9,233,810 | – | 836,031 | 2,771,140 | – | 119 | 6,490 | – | 5,990,064 | 12,011,440 |
| Italy | 2,821,465 | 3,992,743 | 8,481,137 | 415,224 | 673,984 | 2,890,374 | 76 | 252 | 8,198 | 3,236,765 | 4,666,979 | 11,379,709 |
| Japan | – | 4,580,324 | 18,942,261 | – | 639,064 | 6,024,124 | – | 156 | 23,482 | – | 5,219,544 | 24,989,867 |
| Norway | 207,947 | 319,665 | 464,565 | 46,205 | 70,201 | 212,602 | 42 | 78 | 516 | 254,194 | 389,944 | 677,683 |
| Sweden | 493,286 | 731,639 | 1,071,998 | 94,021 | 137,885 | 480,905 | 42 | 73 | 1,242 | 587,349 | 869,597 | 1,554,145 |
| United Kingdom | 3,724,878 | 5,116,813 | 6,976,005 | 481,639 | 1,004,066 | 2,620,986 | 201 | 569 | 8,442 | 4,206,718 | 6,121,448 | 9,605,433 |
| USA | 7,553,659 | 13,961,358 | 25,943,823 | 1,125,265 | 245,061 | 10,543,194 | 2,760 | 5,362 | 48,430 | 8,681,684 | 16,511,781 | 36,535,447 |

Relative frequencies

| Country | Ages 65–79 (per 100) | | | Ages 80–99 (per 100) | | | Ages 100+ (per 1,000) | | | Ages 65+ (per 100) | | |
|---|---|---|---|---|---|---|---|---|---|---|---|---|
| | 1939 | 1960 | 2005 | 1939 | 1960 | 2005 | 1939 | 1960 | 2005 | 1939 | 1960 | 2005 |
| Canada | 5.49 | 6.41 | 9.60 | 0.94 | 1.26 | 3.40 | 0.01 | 0.02 | 0.13 | 6.43 | 7.66 | 13.01 |
| Denmark | 6.60 | 8.90 | 10.93 | 1.07 | 1.61 | 4.07 | 0.00 | 0.00 | 0.12 | 7.66 | 10.51 | 15.01 |
| France | 9.00 | 9.64 | 11.89 | 1.36 | 2.00 | 4.51 | 0.00 | 0.01 | 0.19 | 10.36 | 11.64 | 16.42 |
| Germany | – | – | 14.33 | – | – | 4.20 | – | – | 0.10 | – | – | 18.54 |
| Germany-East | – | 11.72 | 15.32 | – | 1.82 | 4.10 | – | 0.00 | 0.10 | – | 13.54 | 19.43 |

(continued)

**Table 6.2** (continued)

Relative frequencies

| Country | Ages 65–79 (per 100) | | | Ages 80–99 (per 100) | | | Ages 100+ (per 1,000) | | | Ages 65+ (per 100) | | |
|---|---|---|---|---|---|---|---|---|---|---|---|---|
| | 1939 | 1960 | 2005 | 1939 | 1960 | 2005 | 1939 | 1960 | 2005 | 1939 | 1960 | 2005 |
| Germany-West | – | 9.32 | 14.08 | – | 1.51 | 4.22 | – | 0.00 | 0.10 | – | 10.83 | 18.31 |
| Italy | 6.50 | 8.09 | 14.51 | 0.96 | 1.36 | 4.94 | 0.00 | 0.01 | 0.14 | 7.46 | 9.45 | 19.46 |
| Japan | – | 4.97 | 15.02 | – | 0.69 | 4.78 | – | 0.00 | 0.19 | – | 5.66 | 19.82 |
| Norway | 7.06 | 8.96 | 10.09 | 1.57 | 1.97 | 4.62 | 0.01 | 0.02 | 0.11 | 8.63 | 10.93 | 14.71 |
| Sweden | 7.81 | 9.81 | 11.90 | 1.49 | 1.85 | 5.34 | 0.01 | 0.01 | 0.14 | 9.30 | 11.66 | 17.25 |
| United Kingdom | 7.84 | 9.77 | 11.62 | 1.01 | 1.92 | 4.37 | 0.00 | 0.01 | 0.14 | 8.85 | 11.68 | 16.00 |
| USA | 5.79 | 7.82 | 8.79 | 0.86 | 1.42 | 3.57 | 0.02 | 0.03 | 0.16 | 6.66 | 9.24 | 12.38 |

*Source* Human Mortality Database

"–" indicates that data were not available

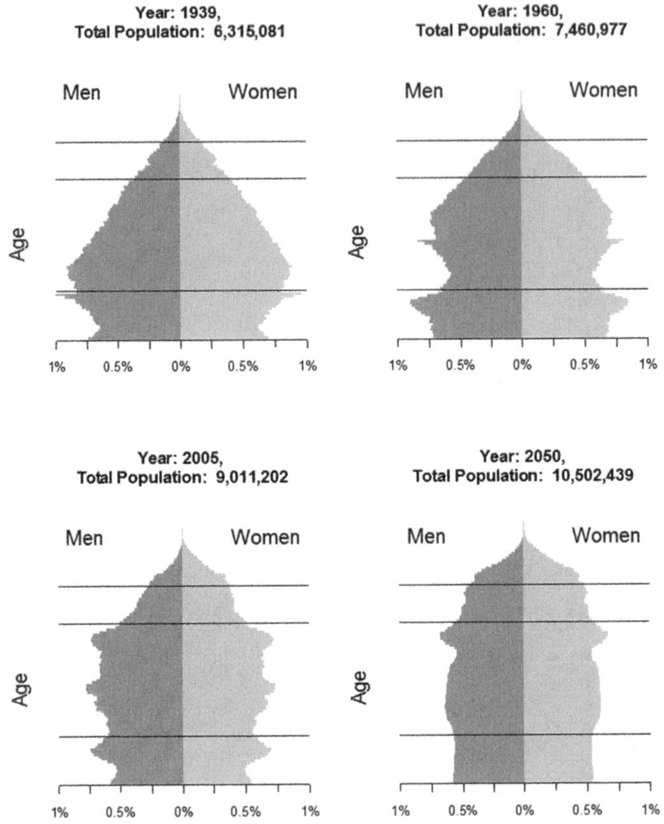


**Fig. 6.1** Population pyramids for Sweden in 1939, 1960, 2005, and 2050. *Source* Human Mortality Database (years 1939, 1960, 2005); Statistics Sweden (year 2050). *Note* Horizontal lines are at ages 20, 65, and 80

ages 20, 65, and 80 to facilitate easy distinction between working ages, retirement ages, and the "oldest-old".

In 1939, the population pyramid for Sweden still largely resembled a "pyramid" with a monotonically decreasing population size by increasing age. But already in that year, we can detect a reduction in the number of children and very young adults as a result of the drastically falling numbers of births since the beginning of the twentieth century.[4] By 1960, the population pyramid was rather onion-shaped with significantly more people aged 65 and older than before. Belonging to the "oldest-old" was, nevertheless, still rather rare. The more vertical shape between ages 65 and 80 in 2005 give some indication that survival chances had improved remarkably

[4] For example, between 1910 and 1935, the number of births in Sweden dropped about 40% from 135,625 to 85,906, according to estimates from the Human Mortality Database (results not shown).

after age 65. If the assumptions of Statistics Sweden are correct, the shape of the population structure will have deviated even further from the shape of a pyramid or onion by the middle of the twenty-first century.

## 6.3 Age-Specific Mortality

The growth in the number of old people and the changing shapes of population pyramids result from age-selective migration, changing fertility patterns, and, most importantly, from declining mortality. In this section, we focus exclusively on trends between 1939 and 2005 in survival and mortality in our selected countries.

Based on period life tables from the Human Mortality Database, we estimate various indicators for the development of longevity for women and men in the countries Jan Hoem lived in for extended periods (Norway, Denmark, Sweden, and East Germany). The second column of Table 6.3 refers to survival until age 65. Hence it is a reflection of mortality *below* this age, whereas the remaining columns illustrate survival after age 65, or age 80.

In 2005, in all countries under study about 90,000 out of 100,000 newborn girls would survive to age 65 under the assumption of unchanged mortality levels at the 2005 values. This number varies from less than 89,000 in Denmark to almost 92,000 in Sweden. In the two other countries, the probability of survival to age 65 reached the value of 0.91. Among males the variation between countries is bigger. The theoretical number of survivors at age 65 from an initial cohort of 100,000 boys is about 81,000–82,600 in the former GDR and Denmark, whereas the situation is more advantageous in Norway and Sweden with corresponding values of 86,000 and 87,000 surviving men.

Once women make it to age 65, under the 2005 mortality regime, about 68 % (Denmark) to 76 % (Sweden and Norway) could expect to become 80 years old and about 2.5 % of the 80-year olds to become centenarians. The situation for men is less favorable: between 56 % (Denmark) and 63 % (Sweden) 65-year old men reach age 80, and less than one percent of these become centenarians.

Nevertheless, we can observe improving chances of survival over time in *all* countries for women *and* men. The probability to survive to age 65 increased in all countries, but it was the improvements in survival between ages 65 and 80 that were truly remarkable. If we consider the lowest and highest probability of surviving from age 65 to 80 among women in our set of countries, these numbers almost doubled in the study period: from a low value of 0.41 in Denmark in 1939 to a high value of 0.76 in Norway and Sweden in 2005. Survival probabilities also increased among men at the same ages, albeit on a smaller scale: between 56 and 63 % of men in our selected countries can expect to survive another 15 years after their 65th birthday.

Although the chances to become a centenarian are still relatively small—even for octogenarians—the situation has dramatically improved. The biggest advance can be observed between 1960 and 1990. With a probability of 2.6 % for octogenarians, the chance for Norwegian women is seven times higher in 2005

**Table 6.3** Selected indicators for survival and longevity of Danish, (East-) German, Norwegian and Swedish women and men in 1939, 1960, 1990, and 2005

| Years | Survivors at age 65[a] | Probability to survive | | Age until X % survive from age 65 | | | |
|---|---|---|---|---|---|---|---|
| | | from 65 to 80 | from 80 to 100 | 90 % | 50 % | 10 % | 1 % |
| *Women* | | | | | | | |
| Denmark | | | | | | | |
| 1939 | 69,855 | 0.414 | 0.00249 | 69/70 | 79/80 | 88/89 | 96/97 |
| 1960 | 81,727 | 0.510 | 0.00329 | 70/71 | 81/82 | 90/91 | 97/98 |
| 1990 | 83,814 | 0.635 | 0.01621 | 71/72 | 84/85 | 94/95 | 101/102 |
| 2005 | 88,623 | 0.678 | 0.02555 | 72/73 | 85/86 | 95/96 | 102/103 |
| Germany (East) | | | | | | | |
| 1960 | 78,639 | 0.462 | 0.00137 | 70/71 | 80/81 | 89/90 | 95/96 |
| 1990 | 84,449 | 0.562 | 0.00543 | 71/72 | 82/83 | 91/92 | 98/99 |
| 2005 | 90,938 | 0.733 | 0.02243 | 74/75 | 86/87 | 95/96 | 102/103 |
| Norway | | | | | | | |
| 1939 | 72,659 | 0.484 | 0.00355 | 70/71 | 80/81 | 90/91 | 97/98 |
| 1960 | 85,215 | 0.556 | 0.00602 | 71/72 | 82/83 | 92/93 | 98/99 |
| 1990 | 88,396 | 0.664 | 0.02043 | 72/73 | 84/85 | 94/95 | 101/102 |
| 2005 | 91,010 | 0.756 | 0.02642 | 74/75 | 87/88 | 96/97 | 102/103 |
| Sweden | | | | | | | |
| 1939 | 70,720 | 0.430 | 0.00131 | 69/70 | 79/80 | 89/90 | 95/96 |
| 1960 | 83,593 | 0.519 | 0.00373 | 71/72 | 81/82 | 90/91 | 97/98 |
| 1990 | 89,375 | 0.699 | 0.01651 | 73/74 | 85/86 | 95/96 | 101/102 |
| 2005 | 91,565 | 0.758 | 0.02604 | 74/75 | 87/88 | 96/97 | 102/103 |
| *Men* | | | | | | | |
| Denmark | | | | | | | |
| 1939 | 65,762 | 0.378 | 0.00161 | 69/70 | 78/79 | 88/89 | 95/96 |
| 1960 | 74,394 | 0.429 | 0.00200 | 69/70 | 79/80 | 89/90 | 96/97 |
| 1990 | 75,426 | 0.440 | 0.00650 | 69/70 | 79/80 | 90/91 | 98/99 |
| 2005 | 82,645 | 0.556 | 0.00926 | 70/71 | 82/83 | 92/93 | 99/100 |
| Germany (East) | | | | | | | |
| 1960 | 67,431 | 0.359 | 0.00074 | 68/69 | 78/79 | 87/88 | 94/95 |
| 1990 | 69,223 | 0.382 | 0.00230 | 69/70 | 78/79 | 88/89 | 95/96 |
| 2005 | 81,012 | 0.564 | 0.00946 | 70/71 | 82/83 | 92/93 | 99/100 |
| Norway | | | | | | | |
| 1939 | 66,366 | 0.434 | 0.00149 | 69/70 | 79/80 | 89/90 | 95/96 |
| 1960 | 75,594 | 0.476 | 0.00408 | 69/70 | 80/81 | 90/91 | 97/98 |
| 1990 | 78,339 | 0.471 | 0.00781 | 69/70 | 80/81 | 91/92 | 98/99 |
| 2005 | 86,177 | 0.614 | 0.00920 | 71/72 | 83/84 | 93/94 | 99/100 |
| Sweden | | | | | | | |
| 1939 | 65,627 | 0.399 | 0.00076 | 69/70 | 78/79 | 88/89 | 95/96 |
| 1960 | 76,107 | 0.434 | 0.00203 | 69/70 | 79/80 | 89/90 | 96/97 |
| 1990 | 81,599 | 0.515 | 0.00662 | 70/71 | 81/82 | 91/92 | 98/99 |
| 2005 | 87,253 | 0.626 | 0.00826 | 71/72 | 83/84 | 93/94 | 99/100 |

[a] Survivors from the radix of 100,000

*Source* Authors' own estimations based on data from the Human Mortality Database

than it was in 1939. The corresponding improvement in Sweden is 20-fold. Also males experienced a sharp increase in survival since the 1960s. Still, their probability to survive an additional 20 years after the 80th birthday is considerably smaller than the probability for women.

While the left side of Table 6.3 depicts various survival indicators *for a given age*, the right side depicts the corresponding quantile functions: which age is attained *for a given fraction of survivors and a given level of survival probabilities in a certain year.*

Assuming a retirement age of 65, in all countries except Denmark, approximately 90 % of women in 2005 will enjoy 10 years of retirement. About 50 % will enjoy more than 20 years of retirement. 10 % will survive to their mid-90s and 1 % will turn 102 years old—if mortality levels remain as those observed in 2005. As expected, male mortality is higher, implying, for instance that less than half of men will be able to live an additional 20 years after retirement at age 65.

## 6.4 Life Expectancy

Life expectancy at birth, $e_0$, as derived from period life tables, remains the most widely used measure to summarize the mortality experience of a population at a given moment of time, despite being criticized recently (for an overview, see Barbi et al. 2008).

During the last 160 years, the world has seen an unprecedented and steady increase in record life expectancy (Oeppen and Vaupel 2002), i.e. the highest life expectancy recorded in a national population in a given calendar year. On average, record life expectancy has been increasing almost linearly by 3 months every year for females. The annual gain in life-years was almost as linear for men as for women, albeit with a weaker slope of 0.222 years per year for males (as compared to 0.243 years for females).

In the two upper panels of Fig. 6.2, we plot with solid black lines the values of the record life expectancy since 1940, for women (left panel) and men (right panel), separately. Until about 1985, countries such as Iceland, the Netherlands, New Zealand, Norway, Sweden, and Switzerland were those with the highest chances of survival for both sexes. During the last 20 years, Japan has been the unchallenged leader in record life expectancy for women, whereas the record for men is shared by Iceland and Japan.

In addition to plotting the life expectancies in the best-performing countries, we also included in the upper panels of Fig. 6.2 the data for our selected four countries. These are depicted by different symbols. The remaining gray lines represent all other countries represented in the Human Mortality Database (approximately 30). This allows us to evaluate the relative performance of our four countries of interest in relation to most other developed countries. From 1940 to 1970, Norway was the country where women and men could expect to live longer than in any other country of the world. Sweden and Denmark lagged slightly behind. In the former GDR,

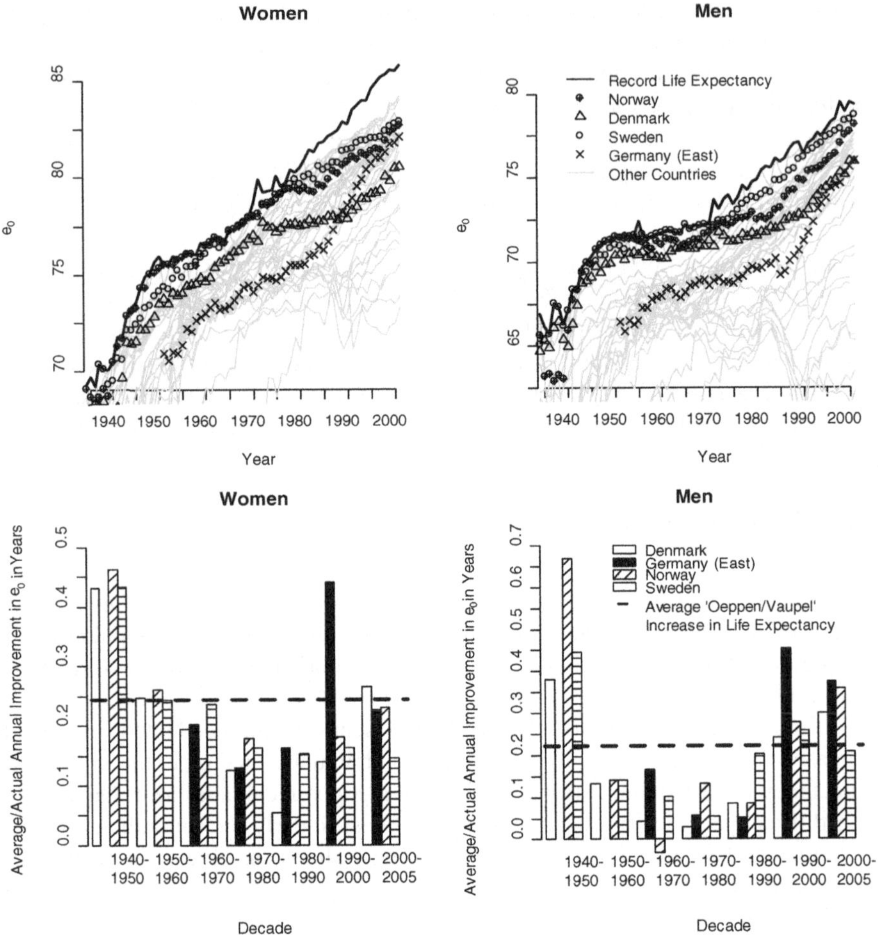


**Fig. 6.2** Life expectancy in selected countries for women and men. Development in life expectancy in Denmark, Germany (*East*), Norway, and Sweden in relation to the record (*Upper Panels*) and average increase (*Lower Panels*) from Oeppen and Vaupel (2002), by decade. *Source* Authors' own estimations based on data from the Human Mortality Database

in 1956 (the first year of available data), female life expectancy was about 3–4.5 years below the corresponding numbers in Scandinavia. For men, the difference was slightly larger (4–5 years). In the next decades, in this group of countries the life expectancy of females kept on increasing steadily only in Sweden. In the early 1980s, the development of female life expectancy in Norway stagnated for a period of time, resulting in Swedish life expectancy over-taking that of their Norwegian peers. Also for men, life expectancy in Sweden has been longer than in Norway since the 1980s. Comparable to Norway, the life expectancy of females in Denmark showed signs of stagnation for some periods. From 1980 until the mid-1990s life expectancy rose by only about half a year, from 77.2 to 77.8 years. It has been argued that the high prevalence of smoking among Danish women was to blame for this period of stagnation, after which Danish life expectancy started increasing again (e.g. Jacobsen et al. 2002; Juel 2000).

While the recent slow improvements in survival among Danish women likely can be attributed to a *cohort effect* of high prevalence of smoking among women born between the two World Wars, the rapid increase in life expectancy after re-unification of the former GDR with West Germany can be explained by *period effects* in mortality (Scholz and Maier 2003; Vaupel et al. 2003). In 1990, life expectancy at birth for East German women was more than 4 years lower than in Sweden (76.3 vs. 80.4 years). Within 15 years after re-unification, this difference became less than one year (81.8 vs. 82.8 years of life expectancy).

Even if lower than among the Norwegians and Swedes, life expectancy among men in the former GDR rose steeply after re-unification as well, to become on the level of Danish men in 2005 (both around 76 years). In 1990, the gap in male life expectancy between former East Germany and Sweden was about 5.6 years. 15 years later, this difference had been cut in half. In contrast to the study of Denmark's mortality developments, it has not yet been conclusively analyzed which factors played the dominant role in reducing post-reunification mortality in the former GDR. Most likely, those factors are less related to personal behavior (like smoking in Denmark) than to structural changes related to standard of living, medical care, etc.

The lower two panels of Fig. 6.2 focus on the average annual improvement in life expectancy in the four countries, by decade, for women (left) and men (right). The dashed horizontal lines represent the expected average annual increase derived from the development of record life expectancy during the last 160 years (Oeppen and Vaupel 2002). Although record life expectancy increased steadily, the trend in mortality improvements is not linear in single countries. As shown already by Oeppen and Vaupel (2002), periods of catching up to record life expectancy are often followed by periods of stagnation and vice versa in single countries.

The 1940s was a decade of rapid improvements in life expectancy for the three Scandinavian countries. Within those 10 years, life expectancy rose by about 4 years for women. Similar values are observed for males in Denmark and Sweden. The largest increase in life expectancy is recorded for Norwegian males: more than 6.5 years between 1940 and 1950.

The 1970s and 1980s were years of poor performance in life expectancy developments for women in all the countries under study. The period of very slow increase was even longer for men (1950–1990). Between 1960 and 1970, the number of expected years of life since birth even decreased for men in Norway.

Indicated by the upper panels, the former East Germany experienced sudden increases in life expectancy since 1990. Norway and Denmark again experienced life expectancy increases, on par with the average increase of record life expectancy. Among males, the additional years of life won were even larger than the reference value of 0.222 years annually.

## 6.5 Inequalities in Life-Span: $e^{\dagger}$

In addition to the average length of life, inequality in lifespan is an important indicator of population health. From the perspective of the individual, it appears to be desirable and related to reduced uncertainty that oneself has approximately the same number of years to live as one's fellow citizens (i.e. low levels of inequality). From a mathematical perspective, though, decreasing inequality in life-span implies smaller possible gains in life expectancy for a given level of survival improvement (see, for instance, Keyfitz and Golini 1975; Vaupel and Canudas-Romo 2003; or Keyfitz and Caswell 2005).

We have measured inequality in lifespan by $e^{\dagger}$ (Vaupel and Canudas-Romo 2003; Zhang and Vaupel 2009)

$$e^{\dagger} = \int_{0}^{\infty} e(a)d(a)da,$$

where $\infty$ denotes the highest attained/attainable age.[5] The enumerations $e(a)$ and $d(a)$ denote remaining life expectancy at age $a$ and numbers of death at age $a$, respectively. Hence, $e^{\dagger}$ is a weighted average. It measures the mean numbers of life-years lost.

In the large panels of Fig. 6.3, we plot annual $e^{\dagger}$ values since 1939 for Denmark, Norway, Sweden, and East Germany, for women (left) and men (right). The small panels illustrate the relationship between $e_0$ on the one hand and $e^{\dagger}$ on the other.

Although the average number of life-years lost ($e^{\dagger}$) is higher for men, the basic trends are the same for both sexes. During the first decades of our observation period, $e^{\dagger}$ was decreasing rapidly. In 1939, women in Scandinavia died on average about 14–15 years before their possible end time. The corresponding numbers for males were about 1–2 years higher. Figure 6.3 shows that the rapid reductions in $e^{\dagger}$ slowed down in the 1970s. By then, the average number of life-years lost was between 10 and 11 years for women and approximately 12 years for males. Analogously to the stagnation in the development of life expectancy, the age-inequalities in mortality did not decrease further among Danish women for several years. By the end of our observation period, the least number of life-years lost was observed in Sweden with about 9.1 years for women and 9.9 years for men. The slow-down in decreasing $e^{\dagger}$ could be interpreted as evidence of people reaching the upper limits of the rectangularized survival curve, i.e. reaching the maximum compression of mortality. This is not inconsistent with continued rises in life expectancy. The distribution of deaths, as measured by the modal (adult) age at death, has been and still is shifting to the right (see, for instance, Canudas-Romo

---

[5] Often denoted as $\omega$ in the demographic literature.

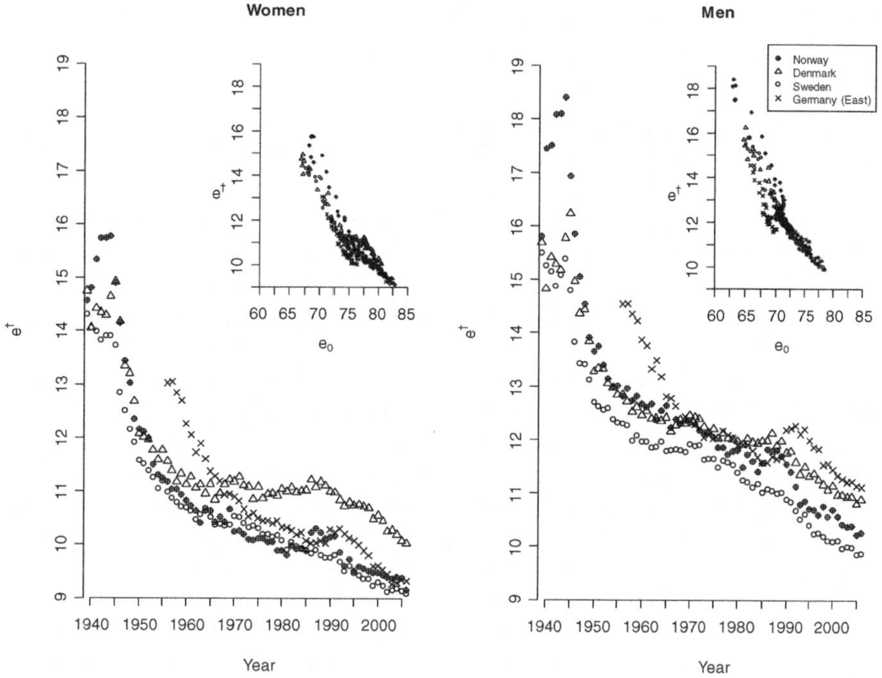


**Fig. 6.3** Average number of life-years lost, $e^{\dagger}$, in Denmark, Norway, Sweden, and Germany (*East*) for females (*left*) and males (*right*) since 1939 (*large panels*). Relationship between $e_0$ and $e^{\dagger}$ for the same countries and the same time period (*small panels*). *Source* Authors' own estimations based on data from the Human Mortality Database

2008; Ouellette and Bourbeau 2009): "The shifting mortality scenario, where the compression of mortality has stopped, may be a realistic description of the current situation in low mortality countries." (Canudas-Romo 2008: 1197–1198).

Each of the two large panels contains a small panel illustrating the strong link between $e_0$ and $e^{\dagger}$. Again, we have depicted our standard set of four countries (as well as the same time horizon); the findings, though, apply to developed countries in general (see Vaupel et al. 2011). Although the relationship is not as striking as the straight increase in record life expectancy over time, there is a strong relationship between $e_0$ and $e^{\dagger}$: The higher the life expectancy, the lower the number of life-years lost. Or expressed differently: Countries where disparities in the age-at-death distribution are relatively small are those where people can enjoy—on average—relatively long lives.

This is less obvious than one might think. As shown in several papers, for instance by Keyfitz and Golini (1975) or Vaupel and Canudas-Romo (2003), with decreasing life-span disparities, mortality levels have to drop at a much faster pace in order to obtain constant increases in life expectancy.

## 6.6 We Live Longer but also Healthier

As life expectancy rose in Europe, the number of years of good health increased too. It is not only the length of life but also its quality that has improved over time. In Table 6.4 we present female life expectancy and healthy life expectancy (HLE) at birth and age 60 during 2000–2002. Healthy life expectancy is defined as the "average number of years that a person can expect to live in *full health* by taking into account years lived in less than full health due to disease and/or injury" (World Health Organization 2004: 96). The WHO definition of disability and health is based on a self-assessed condition-specific state of health as derived from a representative sample of respondents from the Global Burden of Disease study. Due to the lack of data for the former GDR, we include in Table 6.4 instead the value of the indicator for East and West Germany combined.

Between 2000 and 2002, in all the countries under study, the growth in the number of years lived in good health exceeded the development in the number of years lived. For example, in Sweden a newborn girl could expect to live about 82 years; out of these, 72.7 would be years in good health according to the data for 2000. Two years later, she could expect to live to about the same age, and as much as 74.8 years in good health. Similarly, in all the European countries under study, growth in the expected number of years lived in good health for those aged 60 was greater than the increase in life expectancy itself. Over the period studied, Denmark lagged behind in life expectancy and also with regard to the expected number of years lived in good health. The difference between the two numbers—i.e. the expected number of years lived in bad health—was, however, similar in Denmark to the corresponding values in other countries. This means that Danish women spent a larger share of their life-span in bad health than did women in the

**Table 6.4** Female life expectancy and healthy life expectancy (HLE) at birth and age 60, 2000–2002

| Country | Life expectancy | | | Healthy life expectancy | | |
|---|---|---|---|---|---|---|
| | 2000 | 2001 | 2002 | 2000 | 2001 | 2002 |
| *At Birth* | | | | | | |
| Denmark | 79.1 | 79.2 | 79.3 | 70.1 | 70.8 | 71.1 |
| Germany | 81.0 | 81.3 | 81.3 | 71.5 | 72.2 | 74.0 |
| Norway | 81.4 | 81.5 | 81.5 | 72.3 | 72.2 | 73.6 |
| Sweden | 82.0 | 82.1 | 82.1 | 72.7 | 73.2 | 74.8 |
| Japan | 84.6 | 84.9 | 85.2 | 76.3 | 75.8 | 77.7 |
| *At Age 60* | | | | | | |
| Denmark | 22.2 | 22.3 | 22.3 | 16.5 | 16.7 | 17.2 |
| Germany | 23.6 | 23.9 | 23.9 | 17.6 | 17.7 | 19.0 |
| Norway | 24.0 | 24.0 | 24.0 | 18.2 | 17.9 | 18.9 |
| Sweden | 24.3 | 24.3 | 24.3 | 18.7 | 18.5 | 19.6 |
| Japan | 26.8 | 27.1 | 27.4 | 21.4 | 20.7 | 21.7 |

*Sources* WHO, World Health Report, various years; Human Mortality Database

other countries in Table 6.4. According to Jeune and Brønnum-Hansen (2008), however, if one considers slightly longer time perspective (1994–2005), there has been some further increase in the expected number of years lived in good health in Denmark. The authors show that during 1994–2005 life expectancy without long-standing limiting illness increased more than life expectancy during the same calendar period, both for women and men. Thus, the expected life-time with long-standing illness decreased and there was a compression of morbidity to later life. Compared to the corresponding numbers in Japan, there is still plenty of room for improvement in Europe. Japan is a leader not only in life expectancy among women, but also in healthy life expectancy, both at birth and at age 60.

## 6.7 Summary and Discussion

As the oldest-old continent of the world, Europe will continue setting the pace in population aging in the near future. In 2005, between 15 and 20 % of Europeans were 65 years or older. As the simple measure of population counts does not separate the effect of population aging as a consequence of improved chances of survival at old ages from other elements of population dynamics, we focused our attention in this study on European mortality trends, using data on four Northern European countries, Denmark, East Germany, Norway, and Sweden.

Under current mortality regimes, about 90,000 out of 100,000 newborn girls in the countries studied reach age 65, which is approximately 20,000 more than in the late 1930s. 70–75 % of 65-year olds can expect to celebrate their 80th birthday and the chances of becoming a centenarian are also increasing (currently 2.5 % for 80-year olds). These unprecedented improvements in survival translate into exceptional increases in life expectancy. Besides this, we also observe two other positive trends: First, life-span disparities, as measured by $e^{\dagger}$, decreased considerably during our observation period. This means that people die on average fewer and fewer years "before their time". Furthermore, we detected a strong negative linear relationship between life expectancy and $e^{\dagger}$: countries with high life expectancy are also the ones with relatively small life-span inequalities. In addition, people do not only live longer; typically, they also live longer in good health.

We are convinced that our results are not distorted by the impact of poor data. According to an assessment by Jdanov et al. (2008), none of the countries in our analysis have ever had data of poor quality. Since the 1970s, the data for the oldest-old in the countries studied have been of good quality (highest grade, Sweden and Denmark), or at least of acceptable quality (former GDR and Norway).

Are our results typical for Europe as a whole or do they represent an exception from other Western countries? As shown, for instance by Tuljapurkar et al. (2000), Oeppen and Vaupel (2002), and White (2002), life expectancy has been increasing in many developed countries. In the twentieth century alone, record life expectancy rose from 60 to almost 85 years (see supplementary online material of

Oeppen and Vaupel 2002) and keeps on increasing. According to the latest reports, record life expectancy for women in 2008 was 86.05 years (Japanese Ministry of Health, Labor and Welfare 2010). However, trends are not universal. Recent articles point towards diverging trends in mortality among some developed countries (e.g. Meslé and Vallin 2006; Jannsen et al. 2007; Muszyńska and Rau 2009; Staetsky 2009). While many countries follow the pace of the best-practice trend, countries such as the United States or the Netherlands featured smaller than expected gains in life expectancy, in particular among women at older ages. Similarly, our four selected countries show some differences in trends as well. Sweden's increase in longevity closely follows the record pace (as set by Japan). Denmark and the former GDR, on the other hand, deviate from this trend. Smoking is often mentioned as an explanation for slow increases in life expectancy in general (e.g. Jannsen et al. 2007; Wang and Preston 2009). In the case of Denmark, it may be the major explanation (e.g. Jacobsen et al. 2002). The exact reasons for the rapid increase in life expectancy in Eastern Germany after reunification are not yet clearly understood. In an overview, Diehl (2008: 90) writes that "improvements in the health care systems, together with the changes in the economic conditions, have predominantly caused the extension of life expectancy".

## 6.8 Outlook

Our findings indicate that despite some periods of stagnation in mortality reduction in some countries, aging is highly plastic—this is something that has been reported before for human as well as non-human species (Vaupel et al. 2003). In opposition to widespread belief, explicit and implicit, from Aristotle to Lexis (1877) and more recent scholars (e.g. Fries 1980), empirical research for recent decades contradicts the notion of a fixed mortality regime at advanced ages. As shown, for instance, by the path breaking publications of Kannisto in the early and mid-1990s (e.g. Kannisto 1994; Kannisto et al. 1994; Kannisto 1996), by Vaupel (1997) or more recently by Rau et al. (2008), old-age mortality is decreasing, in many countries even at an accelerating pace. Indeed, almost 80 % of the gains in record life expectancy since 1990 are due to improved survival chances at ages 65 and above. The oldest-old (80+) alone produced 41.5 % of these achievements (Christensen et al. 2009).

These improved survival chances—at ages where in the past mortality was considered an immutable constant—are, with no doubt, good news for human individuals and their offspring. If recent trends in survival improvements persist into the future, the majority of newborns in many Western countries can expect to become centenarians, as estimated by Christensen et al. (2009). Hence, what we nowadays call exceptionally long lives will be rather commonplace and no longer the privilege of a few.

At the same time, though, the growing numbers of old people with continuously increasing life-spans pose challenges for societies, such as financing healthcare and old-age pensions. In addition to growing demands on social security, the reductions in the proportion of people at "working ages" and hence the total relative working hours in future populations, imply limitations on the supply side of the social system. In part because of low fertility but mainly due to improved survival chances at high ages, modern population pyramids are changing their shapes to that of an "urn". One of the possible solutions to this situation would be the redistribution of work along the life courses of people in aging countries. For example, as suggested by Vaupel and Loichinger (2006), people may need to work longer at older ages but for fewer hours that is standard now-a-days. This solution may be possible to implement since people not only live longer but also spend increasingly large parts of their lives in good health. Furthermore, there are fewer and fewer jobs of such nature in which people become physically exhausted and for this reason have to retire early in their life course.

**Acknowledgments** We would like to thank Zhang Zhen for allowing us to use his R-code (R Development Core Team 2009) to estimate $e^{\dagger}$, and the late Nancy Vaupel for help with language editing. This contribution benefited greatly from the comments, criticism and suggestions of three anonymous reviewers and the editors.

## References

Barbi, E., Bongaarts J., & Vaupel, J. W. (Eds.) (2008). *How long do we live? Demographic models and reflections on tempo effects*. Demographic Research Monographs 5. Heidelberg: Springer.

Canudas-Romo, V. (2008). The modal age at death and the shifting mortality hypothesis. *Demographic Research, 19*(30), 1179–1204.

Christensen, K., Doblhammer, G., Rau, R., & Vaupel, J. W. (2009). Ageing populations: the challenges ahead. *The Lancet, 374*(9696), 1196–1208.

Diehl, K. (2008). Mögliche Faktoren für die rasche Reduktion der ostdeutschen Übersterblichkeit nach der Wiedervereinigung. *Zeitschrift für Bevölkerungswissenschaft, 33*(1), 89–110.

Fries, J. F. (1980). Aging, natural death, and the compression of morbidity. *The New England Journal of Medicine, 303*(3), 130–135.

Jacobsen, R., Keiding, N., & Lynge, E. (2002). Long term mortality trends behind low life expectancy of Danish women. *Journal of Epidemiology and Community Health, 56*(3), 205–208.

Jannsen, F., Kunst, A., & Mackenbach, J. (2007). Variations in the pace of old-age mortality decline in seven European countries, 1950–1999: the role of smoking and other factors earlier in life. *European Journal of Population, 23*(2), 171–188.

Japanese Ministry of Health, Labour and Welfare (2010). Abridged Life Tables for 2008. Available online at http://www.mhlw.go.jp. Downloaded on 26 Jan 2010.

Jdanov, D. A., Jasilionis, D., Soroko, E. L., Rau, R., & Vaupel, J. W. (2008). Beyond the Kannisto-Thatcher database on old age mortality: An assessment of data quality at advanced ages. MPIDR Working Paper WP 2008–2013. Rostock: Max Planck Institute for Demographic Research.

Jeune, B., & Brønnum-Hansen, H. (2008). Trends in health expectancy at age 65 for various health indicators, 1987–2005, Denmark. *European Journal of Ageing, 5*(4), 279–285.

Juel, K. (2000). Increased mortality among Danish women: population based register study. *British Medical Journal, 321*(7257), 349–350.

Kannisto, V. (1994). *Development of oldest-old mortality, 1950–1990: Evidence from 28 developed countries*. Monographs on Population Aging 1. Odense: Odense University Press.

Kannisto, V., Lauritsen, J., Thatcher, A. R., & Vaupel, J. W. (1994). Reductions in mortality at advanced ages: several decades of evidence from 27 countries. *Population and Development Review, 20*(4), 793–810.

Kannisto, V. (1996). *The advancing frontier of survival*. Odense: Odense University Press.

Keyfitz, N., & Golini, A. (1975). Mortality comparisons: the male-female ratio. *Genus 31*(1–4), 1–34 (reprinted in 2009 in *Genus 65*(1), 51–81).

Keyfitz, N., & Caswell, H. (2005). *Applied mathematical demography* (3rd ed.). Berlin: Springer.

Lexis, W. (1877). *Zur Theorie der Massenerscheinungen in der menschlichen Gesellschaft*. Fr. Wagner'sche Buchhandlung.

Meslé, F., & Vallin, J. (2006). Diverging trends in female old-age mortality: The United States and the Netherlands versus France and Japan. *Population and Development Review, 32*(1), 123–145.

Muszyńska, M. M., & Rau, R. (2009). Falling short of highest life expectancy: How many Americans might have been alive in the twentieth century. *Population and Development Review, 35*(3), 585–603.

Oeppen, J., & Vaupel, J. W. (2002). Broken limits to life expectancy. *Science, 296*(5570), 1029–1031.

Ouellette, N., & Bourbeau, R. (2009). Changes in age-at-death distribution in low-mortality countries: A nonparametric approach. Paper presented at the XXVI IUSSP International Population Conference, Marrakech, Sep 27–Oct 2.

Preston, S. H., Himes, C., & Eggers, M. (1989). Demographic conditions responsible for population aging. *Demography, 26*(4), 691–704.

Rau, R., Soroko, E., Jasilionis, D., & Vaupel, J. W. (2008). Continued reductions in mortality at advanced ages. *Population and Development Review, 34*(4), 747–768.

R Development Core Team (2009). R: A language and environment for statistical computing. Vienna, Austria: R Foundation for Statistical Computing.

Scholz, R. D., & Maier, H. (2003). German unification and the plasticity of mortality at older ages. MPIDR WP-2003–2031. Rostock: Max Planck Institute for Demographic Research.

Staetsky, L. (2009). Diverging trends in female old-age mortality: A reappraisal. *Demographic Research, 21*(30), 885–914.

Tuljapurkar, S., Li, N., & Boe, C. (2000). A universal pattern of mortality decline in the G7 countries. *Nature, 405*(6788), 789–792.

United Nations (1968). The concept of a stable population. Application to the study of populations with incomplete demographic statistics. Population Studies 39. New York: United Nations. Department of Economic and Social Affairs.

United Nations (2009). Population Division of the Department of Economic and Social Affairs of the United Nations Secretariat. World Population Prospects: The 2008 Revision and World Urbanization Prospects: The 2007 Revision. Available online http://esa.un.org/unpp, downloaded on 11 Aug 2009.

Vaupel, J. W. (1997). The remarkable improvements in survival at older ages. *Philosophical Transactions of the Royal Society of London: Biological Sciences, 352*(1363), 1799–1804.

Vaupel, J. W., & Canudas-Romo, V. (2003). Decomposing change in life expectancy: A bouquet of formulas in honor of Nathan Keyfitz's 90th birthday. *Demography, 40*(2), 201–216.

Vaupel, J. W., Carey, J. R., & Christensen, K. (2003). It's never too late. *Science, 301*(5640), 1679–1681.

Vaupel, J. W., & Loichinger, E. (2006). Redistributing work in aging Europe. *Science, 312*(5782), 1911–1913.

Vaupel, J. W., Zhang, Z., & van Raalte, A. (2011). Life expectancy and disparity: an international comparison of life table data. *BMJ Open, 1*(1), e000128–e000128.

Wang, H., & Preston, S. H. (2009). Forecasting United States mortality using cohort smoking histories. *PNAS, 106*(2), 393–398.

White, K. M. (2002). Longevity advances in high-income countries, 1955–1996. *Population and Development Review, 28*(1), 59–76.

World Health Organization (2004). *World Health Report*. Geneva: WHO.

Zhang, Z., & Vaupel, J. W. (2009). The age separating early deaths from late deaths. *Demographic Research, 20*(21), 721–730.

# Chapter 7
# Inequalities in Life Expectancy Between and Within European Countries

**Jacques Vallin**

## 7.1 Introduction

*Inequalities in life expectancy between and within European countries* opens up a wide field for questions, research, and interpretative analysis.[1] Inequalities in human well-being are one of the major features of our world. Even if a distinction can be made between inequality, often considered as largely unavoidable and sometimes even necessary to the functioning of societies, and inequity, which is unacceptable and should be eliminated by appropriate policies, some inequalities are judged as more inequitable than others. This is the case for inequality in death. Even if, very probably, perfect equality in length of life could never be achieved between individuals because of a certain number of irreducible biological differences, it is generally considered that inequalities among populations can, for their part, be reduced, if not eradicated, by appropriate policies. Nowadays, there is broad consensus among political institutions and public opinion on this matter, since such differences have been identified and fully acknowledged. But, as yet, recognizing inequalities in life expectancy and fighting them by policies and programmes has proved to be far from sufficient to eliminate them. Even within a rather homogeneous region of industrialized countries like Europe, huge inequalities still exist, both between countries and within countries.

[1] On the occasion of the Symposium *The Demography of Europe* organized in honor of Professor Jan M. Hoem, I was invited to make a presentation on this topic. This chapter is based on this presentation.

J. Vallin (✉)
Emeritus Research Director, Institut National d'Etudes Démographiques (INED), Paris, France
e-mail: vallin@ined.fr

G. Neyer et al. (eds.), *The Demography of Europe*,
DOI: 10.1007/978-90-481-8978-6_7,

Indeed the matter is quite complex and an entire book would not be enough to fully cover the various aspects of life expectancy differentials. Without any claim to do so, the first part of this paper will be devoted to describing some of the major features of the phenomenon. Even more difficult and certainly without any hope of providing definitive evidence, the second part will be an attempt to explain the current diversity of situations.

## 7.2 Describing Diversity

A first level of life expectancy inequalities emerges when comparing countries, and differences arise from various aspects of national mortality patterns and trends, some of which will be mentioned in Sect. 7.2.1, while Sect. 7.2.2 will show a few examples of sub-national diversity.

### *7.2.1 Diversity Between Countries*

In spite of their economic, social, demographic and cultural similarities, European countries show a broad range of life expectancy levels, largely due to the fact that trends observed in the recent decades resulted in a considerable divergence between East and West. But such divergence is not only a matter of levels achieved, it is also due to diverging changes in the structure of mortality.

#### Level Diversity

According to data of the Council of Europe, in 2004, life expectancy at birth, for both sexes, ranged from 65.6 years in Russia to 81.3 years in Iceland. For the 46 countries covered (including Turkey and Caucasian countries), the mean is 75.6 years. Only five countries, however, (Slovakia, Poland, Czech Republic, Slovenia, and Denmark) recorded levels within the range of 74–78 years, and standard deviation is rather high: 11.2 years. This reflects the dramatic gap observed today between Eastern and Western Europe. It is quite surprising to realise that all four western countries of the former USSR (Belarus, Moldavia, Russia, and Ukraine), where the expectation of life ranges between 65 and 69 years, have lower life expectancies than Turkey or the Caucasian countries.[2] At the other end of the spectrum, the most advanced countries are the Netherlands and Iceland with more than 81 years of life expectancy. In fact, according to Fig. 7.1, the

[2] Estimates from current statistics in these countries are not as reliable as for the others. However, the estimates given by the Council of Europe for Armenia and Georgia for 2004 are not far from the results of a very recent deeper analysis (Duthé et al. 2009). In Turkey, rather good data are provided by surveys.

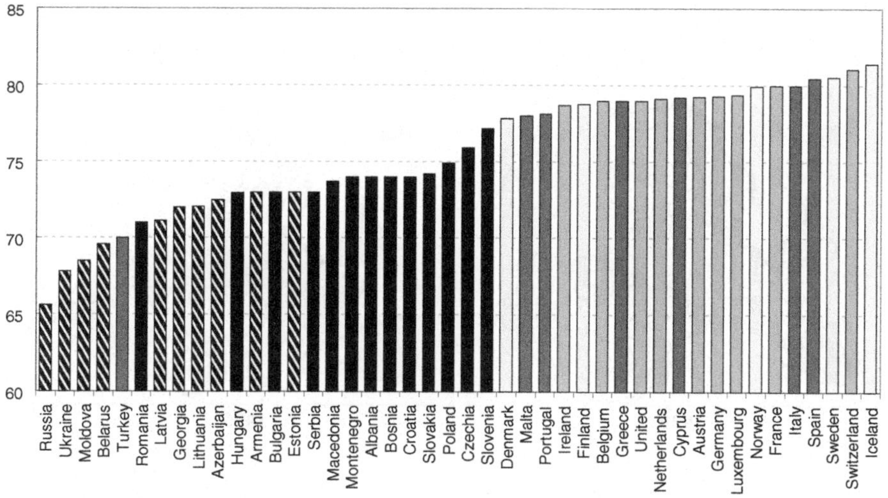


**Fig. 7.1** Life expectancy in Europe in 2004. *Source* Council of Europe (2006)

spectrum is clearly divided into three rather homogeneous groups: the former USSR countries, the other former Communist countries, and all the other countries, which comprise the Mediterranean countries, the Western European countries, and the Nordic countries. The first group is the least homogeneous, with life expectancy ranging from 65 to 73 years, largely because the Baltic countries with about 73 years of life expectancy overlap with the other former Communist countries. Most of the latter fit into a very short interval, between 73 and 74 years of life expectancy, but the three most advanced countries, namely Poland, the Czech Republic, and Slovenia, make a clear transition towards countries belonging to the upper range of life expectancy in Europe. By contrast, this upper category is no longer well divided into the classical distinction between Mediterranean, Western and Nordic countries. The very narrow range between 78 and 80 years of life expectancy includes most western countries except one but also several Mediterranean and Nordic ones, while the three less advantaged countries in this group include one Nordic (Denmark) and two Mediterranean ones (Malta, Portugal). At the same time, the four most advantaged countries include one Mediterranean (Spain), one Western (Switzerland) and two Nordic ones (Sweden, Iceland).

The present clear break between Eastern and Western Europe results from a radical change that has occurred during the last four decades. Figure 7.2 has been drawn in such a way that the ranges of life expectancy remain the same from period to period, so that one can observe the changes in life expectancy across Europe over the period 1950–2004. Thanks to general progress, life expectancy in the ranges of 66.3–73.3, which were totally absent in 1950, occupy a very large surface in 2004, while life expectancy below 59.3 years or between 59.3 and 62.7 years, widespread in 1950, have disappeared in 2004. However the most impressive fact is the change in the way Europe is divided.

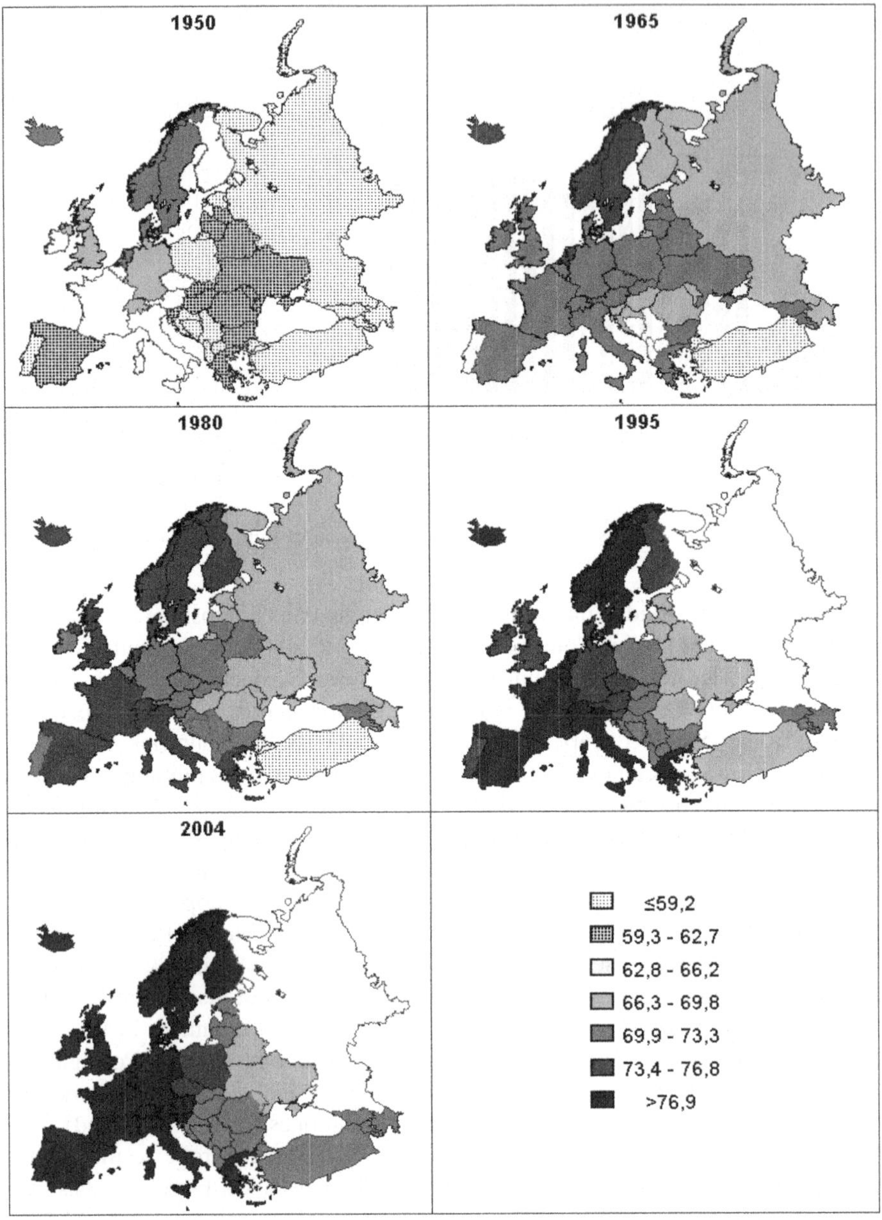


**Fig. 7.2** The changing geography of life expectancy in Europe (1950–2004). *Source* Council of Europe (2006), own calculations

In 1950, there was a strong contrast between a typical northern cone including the four Scandinavian countries (Iceland, Denmark, Norway, Sweden) and the Netherlands, and a southern-eastern half circle from Portugal to Russia.

Fifteen years later, in 1965, this contrast had almost disappeared after a huge movement of convergence that resulted in a much more homogeneous map largely ranging between 69.9–73.3 and 66.3–66.8, close to the mean. The only major contrast is the opposition between Turkey in the South (with life expectancy still below 59.3 years) and the Nordic cone (except Denmark), with a life expectancy of already 73.4–76.8 years. Then, by 1980, the Mediterranean countries join the Nordic ones with a life expectancy of 73.4–76.8. By 1995 all Western European countries have a life expectancy within this range, and the Mediterranean countries as well as the Nordic countries (except Denmark, Finland, and Portugal) even have a life expectancy above 76.8. By contrast, between 1980 and 1995 life-expectancy in Russia and in the Ukraine, Belarus, and Moldova dropped, so that by 1995 these countries belonged to groups with lower life expectancies than they did in 1980. The other former Communist countries retained their life expectancy or increased it to the range of 69.9–73.3, with the remarkable exception of Poland and the Czech Republic that finally joined the group of countries with a life expectancy of 73.4–76.8. The contrast between East and West, quite perfect in 1995, is broken by these two exceptions (Poland and Czech Republic) that are quite significant as we shall see later.

## Structural Diversity: Age Patterns

Furthermore, life expectancy is only a summary of life tables built to combine age-specific mortality rates. The diversity of life expectancy levels may conceal additional diversity in terms of the age structure of mortality, which may or may not be linked to the level of diversity.

In a previous work published a few years ago with France Meslé (Meslé and Vallin 2002), a hierarchical analysis was applied to a set of 28 European countries[3] with a population of at least one million for which we were able to collect statistics on the probability of dying by five-year age groups, for periods of several years around 1965 and 1995. Figure 7.3 illustrates the results obtained for these two periods centred on the same years as the second (1965) and the fourth map (1995) of Fig. 7.2.

In 1965, the situations for both males and females varied sufficiently to establish a distinction between several groups of countries in the dendrograms (i.e. tree diagrams), but no clear contrasts emerge. One group does seem to stand slightly apart from the rest, although it is made up of countries which are not all

[3] Austria, Belgium, Bulgaria, Czech Republic, Denmark, Estonia, Finland, France, East Germany (GDR), West Germany (FRG), Greece, Hungary, Ireland, Italy, Latvia, Lithuania, Netherlands, Norway, Poland, Portugal, Romania, Russia, Spain, Sweden, Switzerland, Ukraine, United Kingdom, and Yugoslavia.

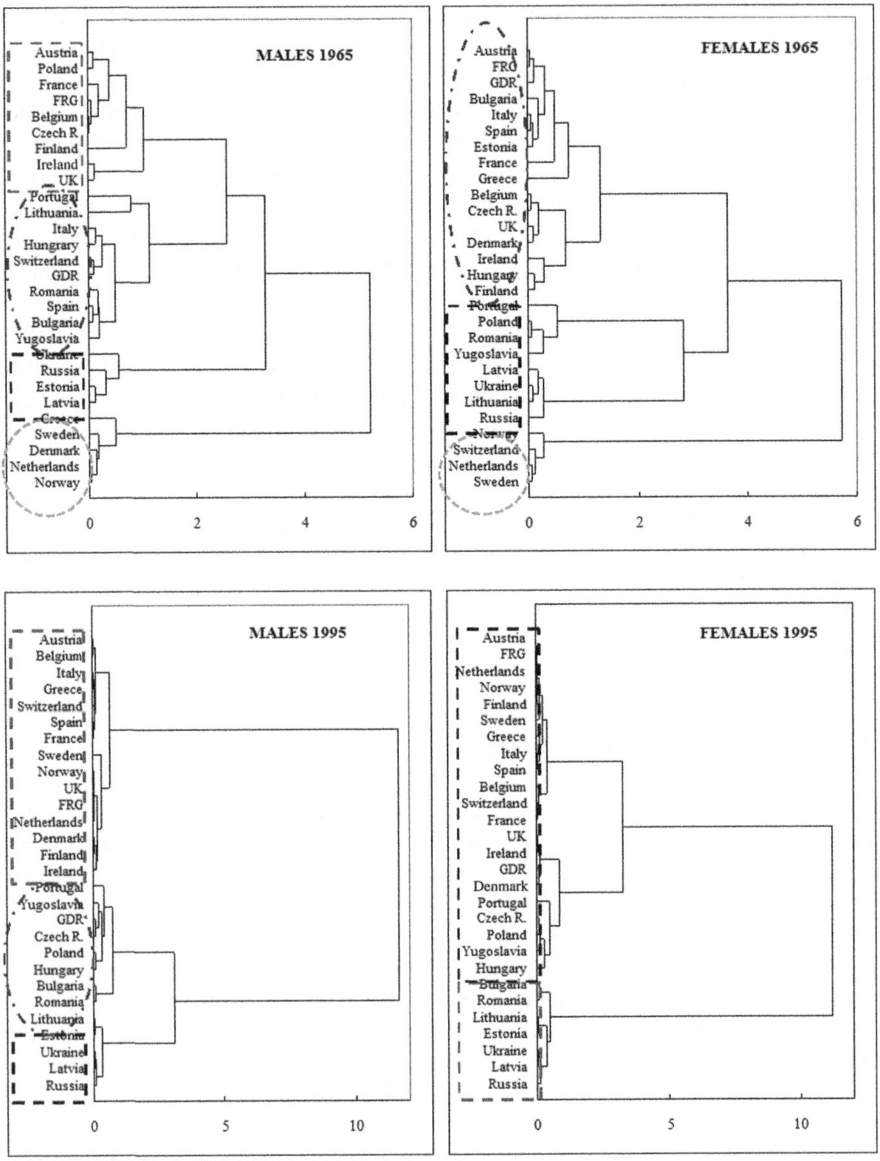


**Fig. 7.3** Summary dendrogram for age-specific probabilities of dying in 28 European countries, in 1965 and 1995. *Source* Meslé and Vallin (2002)

geographically close to one another (Denmark, Greece, Netherlands, Norway and Sweden for males; Netherlands, Norway, Sweden and Switzerland, for females); moreover, the distance between the first and the second node is not much longer than that separating the two following nodes, and this pattern is repeated up to the last branches of the tree. In fact, starting from the furthest node, one can increase the number of partitions at each level without reaching a stage where a final partition becomes obvious. A varying number of groups of countries can be isolated in this manner (possibly four for males and three for females, as underlined on Fig. 7.3), but without bringing out either any sharp contrast between them or any particular geographical logic.

On the contrary, in 1995, the dendrograms show a striking contrast between two or three groups of countries. Among men, the partition into three groups clearly separates, within the Eastern European group, the former USSR (Estonia, Latvia, Lithuania, Russia, Ukraine) from the other former Communist countries (Bulgaria, Hungary, Poland, GDR, Czech Republic, Romania, Yugoslavia), to which one must add Portugal. The case of Portugal is, in a way, an exception confirming the rule. Indeed, as will be seen below, the main factor of divergence in the mortality patterns corresponds to the modification of death risks at adult ages and, in this respect, the situation of Portugal is rather atypical for Western Europe, since the mortality rate among young adults is abnormally high compared to that of other age groups. In order to divide the third group of males into two sub-sets, one has to go very far down in the dendrogram, where two rather large, geographically well-defined groups can be found: a northern group, running from Western Germany to the British Isles and the Scandinavian countries, and a group of Mediterranean and Alpine countries, plus Belgium.

Among females, the divisions suggested by the 1995 dendogram are slightly different. With three groups, the distances are quite long but the geographical continuity is not as marked. Central Europe does contrast with most Western countries but it remains associated with Portugal, the United Kingdom, Ireland and Denmark.

In fact, as we know, the differences in mortality have traditionally been greater among males than among females. Therefore, we compare four groups of countries resulting from the hierarchy among males in 1995, mainly to look at the radical contrast between the two first groups, East and West, but also at the differences shown by the two sets of sub-groups, namely:

- "Mediterranean and Alpine Europe": Austria, Belgium, Spain, France, Greece, Italy and Switzerland (the inclusion of Belgium does not destroy the group's geographical continuity);
- "Northern Europe": Denmark, Finland, Ireland, Norway, Netherlands, West Germany, United Kingdom, and Sweden
- "Central Europe": Bulgaria, Hungary, Poland, East Germany, Czech Republic, Romania and Yugoslavia (removing Portugal in order to preserve the geographical continuity of the group)
- "former USSR": Estonia, Latvia, Lithuania, Russia and the Ukraine.

Figure 7.4 displays the ratio of the 1995 value of each age-specific death rate to its value in 1965, in the four groups.

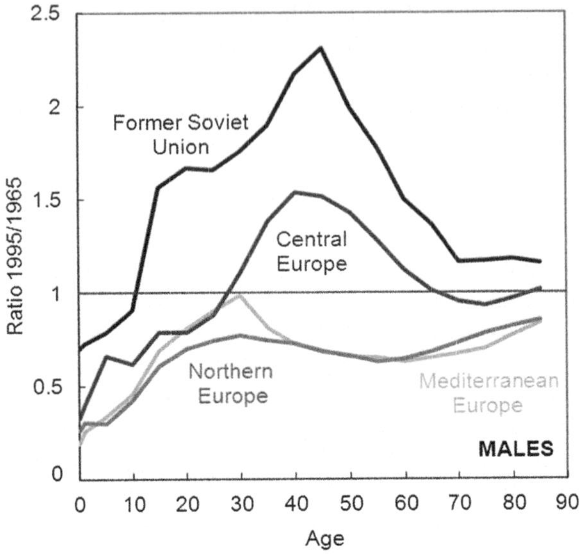


**Fig. 7.4** Relative change in age-specific mortality patterns in four groups of European countries, 1965–1995. *Source* Meslé and Vallin (2002)

The contrast between countries of the former USSR and Western countries is striking, while differences within the latter group are minute, as indicated by the short distance in the two sub-groups shown by the dendrogram. The Central European countries lie clearly in between. Thus, the East–West division of Europe is not only a matter of life expectancy level but also a question of age-specific patterns of mortality. During the past 4 decades, former USSR countries and, to a less dramatic extent, Central European countries were hit by a tremendous increase in male adult mortality while they continued to make significant progress in reducing infant mortality and while old-age mortality remained relatively stable. Western countries, on the contrary, succeeded in reducing mortality at all ages (except around age 30 in the Mediterranean group). Not only have levels of life expectancy diverged sharply since the mid sixties as shown in the maps of Fig. 7.2, but there has also been a striking distortion of the mortality age pattern of the former Communist countries while that of Western countries remains practically unchanged.

## Structural Diversity: Age-and-Cause Patterns

What is true for age patterns is even more obvious for the cause of death structure of mortality. To vary the angle of view, let us consider the best observed life expectancy, the historical trends of which have been discussed recently (Oeppen and Vaupel 2002; Vallin and Meslé 2009). Does the maximum life expectancy observed at one moment in history fully reflect the most ideal mortality pattern? Certainly not. To achieve this objective it would be necessary for the country that enjoys the highest life expectancy to also be the one which enjoys the lowest mortality rates at each age and, even more, the lowest mortality rates at each age for each cause. In a recent paper (Vallin and Meslé 2009) the world's lowest age-specific and age-and-cause specific mortality rates were combined in hypothetical life tables for each year since 1950. Figure 7.5 compares the resulting life expectancies to the maximum life expectancy actually observed.

For females, the combination of the lowest age-specific mortality rates ($e_0$ min-[A]) does not differ significantly from the maximum actually observed first in the Netherlands (from 1950 to 1965) then in Sweden (1966–1980) and finally in Japan (since 1981). This means that nowadays, the variance in mortality age patterns has become very low among the most advanced countries. The highest life expectancy country is also the one where almost all age-specific mortality rates are the lowest. However, this is not yet completely true. Around the beginning of the 1980s, when Japan replaced Sweden as the top-performing country for life expectancy, observed maximum life expectancy was slightly lower than $e_0$ min-[A] because the latter results from the combination of at least two different mortality age patterns, that of Japan, which did not yet surpass Sweden for every age-specific mortality rate, and that of Sweden, which did still better than Japan at some ages (not to mention other countries that may have done better than both Sweden and Japan for some ages). Age-specific patterns of mortality matter slightly more for males, for which a non-negligible gap between observed maximum life expectancy and $e_0$ min-[A] is shown in Fig. 7.5 throughout the 1960s and 1970s. But for both sexes, this gap is practically non-existent when compared to the distance between observed maximum life expectancy and the hypothetical one resulting from the combination of the minimum age-and-cause specific mortality rates ($e_0$ min-[A*C]). When using 12 groups of causes, this distance ranges between 3 and 6 years of life expectancy depending on which sex and period one considers,

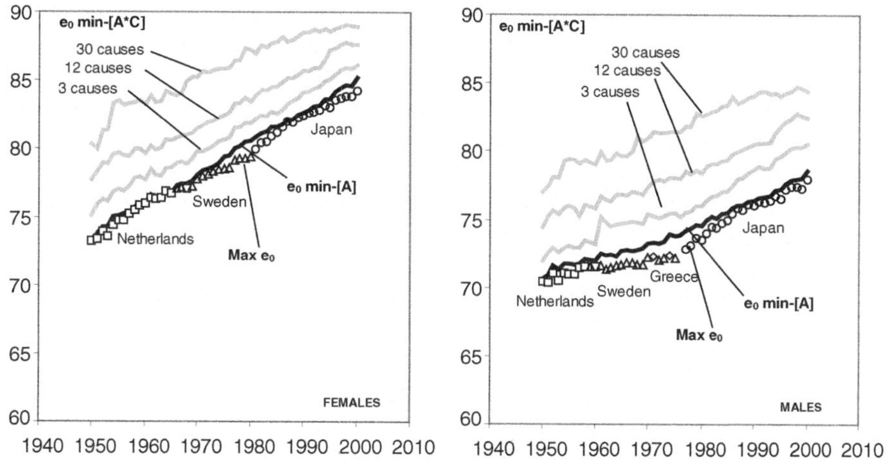


**Fig. 7.5** Highest observed life expectancy (Max $e_0$) compared with different combinations of minimum age or age-and-cause specific rates ($e_0$ min-[A] and $e_0$ min-[A*C]), 1950–2000. *Source* Vallin and Meslé (2008)

instead of mostly less than 1 when $e_0$ min-[A] is considered. Even at the extreme level of the highest life expectancy, mortality structure is an important factor of mortality diversity, especially when causes of death are considered. And, naturally, the larger the number of cause-of-death categories used, the greater is the observed diversity, as shown in Fig. 7.5, comparing three levels of cause-of-death groupings (groups of 3, 12 and 30 causes).

### *7.2.2 Diversity Within Countries*

Since mortality is related to many economic, social, cultural and political factors, one might think that there is less diversity within countries than between countries. At least one of the major factors, the enormous economic difference existing between countries (GDP per capita is about 500 times higher in the world's richest countries than in the poorest), disappears when looking inside a specific country. But also, one might assume that countries are more homogeneous in terms of culture and behaviour, and that government action is able to reduce the effects of many factors of inequality through education, social policies, territorial equilibrium and, especially, health policy and health care systems. In reality, the causes of diversity are no fewer than the factors of homogeneity. Very likely it could be the reverse. Indeed there is only one way to distinguish countries (the borders), while sub-populations can be defined on the basis of many different criteria. Not only can sub-territories be compared, an approach to internal diversity that looks very similar to the cross country analysis, but many other criteria can be used to measure, for example, differences in life expectancy by sex, marital status, level of education, employment, occupation, social class, housing, nationality, religion, not to mention body mass index, cigarette smoking, alcohol consumption or even birth weight, length of gestation, genetic features, etc. And, naturally, various criteria can be combined to obtain increasingly homogeneous categories that will be increasingly different from other categories. Ultimately, this could be refined down to the level of differences between individual life expectancies (if life expectancy were not by definition a statistical mean). According to its discriminatory power, one or another criterion will lead to mortality diversity within a country that is sometimes higher than the level of international diversity, especially if several criteria are combined. For example, a linkage study (Jasilionis et al. 2007) based on four criteria (education, marital status, nationality, and urban/rural) showed that in 2001–2004, Lithuanian male life expectancy at age 30 differed by 11.2 years between married and widowed men (the most discriminatory criterion), but when the four criteria are combined, the difference between a high-educated, married, Lithuanian, urban man and a low-educated, unmarried, non-Lithuanian rural one went up to 27.5 years. This is much more than the distance between Japan and Russia (16.5 years).

Describing all possible life expectancy differences within countries is not possible here. Let us have a modest look at three of them, according to geography, sex, and socio-cultural groups.

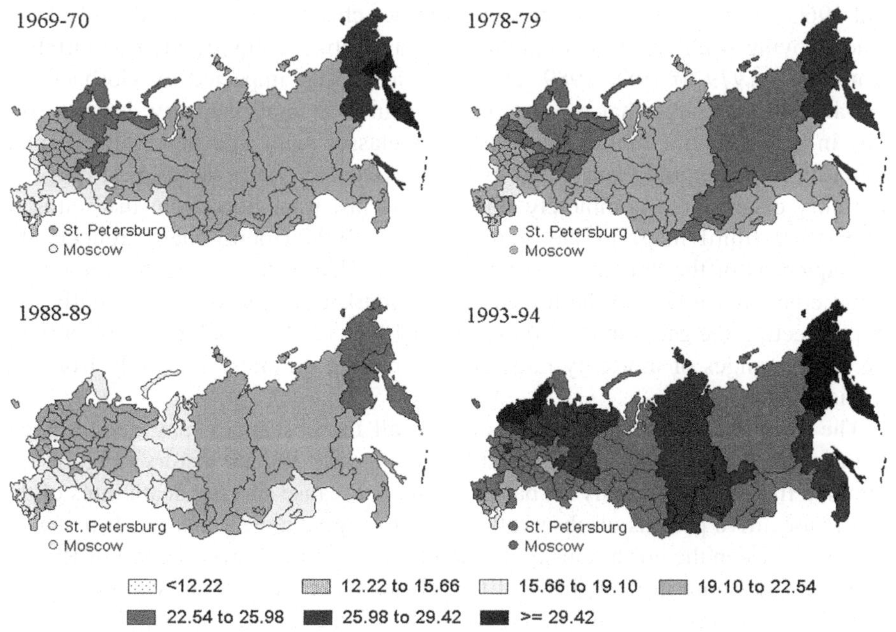


**Fig. 7.6** Age-standardized male mortality rates among 73 administrative units in Russia in 1969–70, 1978–79, 1988–89 and 1993–94. *Source* Vallin et al. (2005)

## Geography

Analysing geographic variations in mortality within a specific country is not very different from doing international comparisons. Everything said above could apply here. To avoid repetition and taking into account the greater comparability of data produced by the same statistical institution, I will mention here just one example of a more complex analysis that can be done by referring to the results of a collaboration between INED, the MPIDR and the Moscow Center for Demography and Human Ecology (Vallin et al. 2005). The study starts with four maps of the Russian geography of the variation in age-standardized male mortality for four periods centred on the last four censuses then available[4] (Fig. 7.6). All four maps were drawn according to the same seven value classes of mortality levels.[5]

Changes from period to period are influenced by very different types of events that resulted in dramatic fluctuations in Russian life expectancy. From 1969–1970 to 1978–1979, the latter was slowly declining, as it had been doing since the

[4] Not including the most recent one that came later, but including the so-called micro census of 1994.

[5] The size of these classes is of one standard deviation, and the central one is centred on the mean all-Russia value for the four periods mixed together. In fact, since the range of values is much wider above than below this mean value, the central class is the third one.

mid-1960s. The geography of mortality did not change much. We see only large zones moving from one class to the next one reflecting a slightly higher mortality. From 1978–1979 to 1988–1989, the map was mostly impacted by Gorbachev's anti-alcohol campaign that resulted not only in a general mortality decrease but also in a substantial homogenisation. The classes with age-standardized male mortality rates ranging from 15.66 to 25.98 expanded, while the highest range of mortality disappeared completely. Finally, the last transition, from the status in 1988–1989 (third map) to the status in 1993–1994 (fourth map), reflects the consequences of the dramatic mortality increase due to the severe economic and social crisis that followed the transition to a market economy. Consequently, the map reflecting the geography of mortality in 1993–1994 is much more covered by the higher ranges of mortality rates and the pattern differs significantly from the previous ones.

The question arises whether, in spite of all these changes, any geographical features are strong enough to resist such a hectic recent history. Figure 7.7, obtained from a cluster analysis, based on age-and-cause specific death rates of the four consecutive periods of observation, seems to give a clear answer. It reveals a contrast between the north (cluster 4) and the south of the European part of Russia. However, the south itself is made of two clusters, the western and central regions (cluster 1), and the southern rim closer to the Caucasus and Caspian Sea (cluster 2), but both of them extend into the Asian part (extreme south-west of Siberia), which roughly follows the fertile black soil or *chernozem* belt.

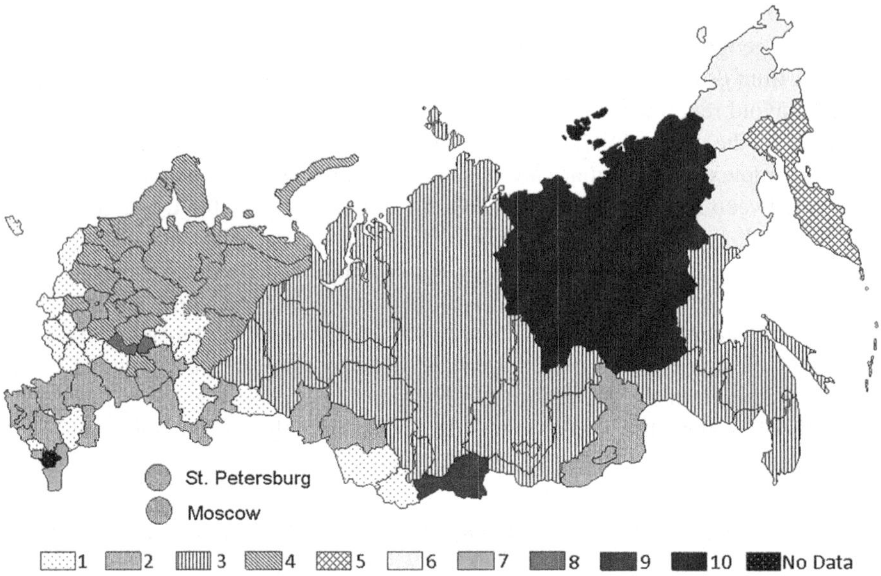


**Fig. 7.7** Global clustering of the four period age-and-cause specific mortality rates (periods: 1969–70, 1978–79, 1988–89, 1993–94). *Source* Vallin et al. (2005)

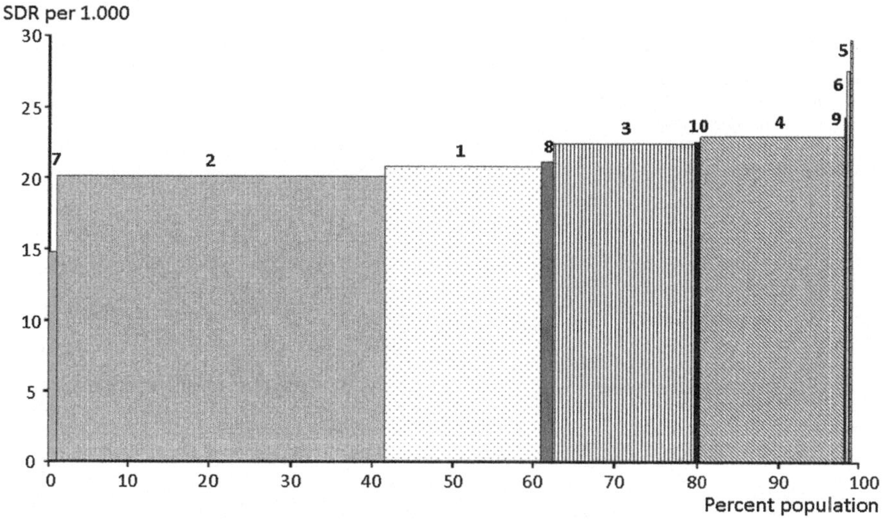


**Fig. 7.8** Distribution of Russian male population by the 10 clusters, ranked by age-standardized death rate (four-period mean). *Source* Vallin et al. (2005)

The fourth important cluster (cluster 3) includes most of the western part of Siberia (from the Urals to Lake Baikal) with a southern extension towards Vladivostok and the Pacific Ocean. The other 6 clusters represent very specific and isolated cases like Dagestan (cluster 7) or Yakutia (cluster 10). Figure 7.8 shows that the first four clusters include an overwhelming share of the population, while the other 6 are very sparsely populated. But it also shows that the differences in mean levels of mortality observed over the four periods are not very large between the main clusters while most of small clusters lie at the extremes. The differences between the main clusters are due more to mortality patterns and changes than to mortality levels.

Indeed, Fig. 7.9 shows important differences in cause-of-death patterns between the four main clusters, as compared to Russia as a whole: For example, mortality from heart diseases, influenza/pneumonia and violent deaths is much higher in cluster 3 (Western Siberia), while chronic respiratory diseases have a much stronger impact in cluster 1 (Northern Chernozem belt) and accidental poisoning by alcohol is more characteristic of cluster 4 (North European Russia). Cluster 1 (Southern Chernozem belt) is mainly distinguished by lower mortality from alcohol poisoning.

## Sex: When Geography Contradicts History

Sex differences in mortality are a fascinating topic, especially when approached from the dual viewpoints of geographic variation and historical trends. In a country like France (Vallin 2002), but also in various other European countries (Vallin

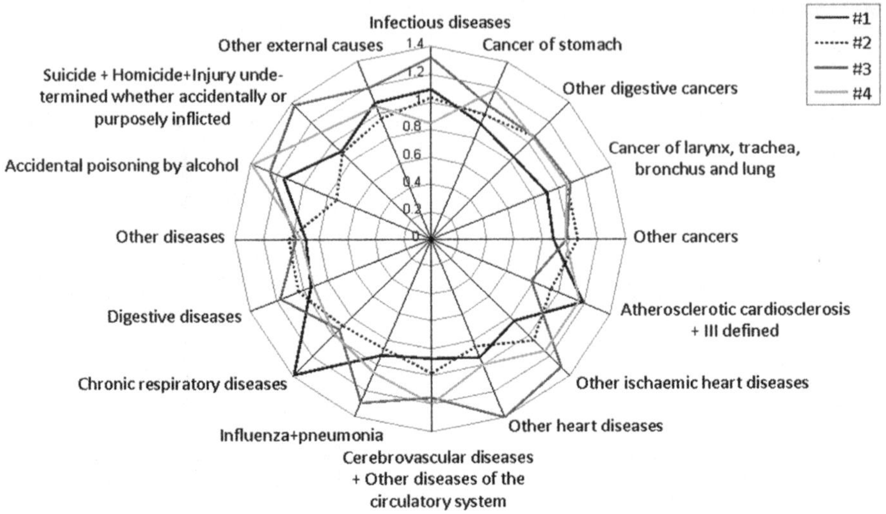


**Fig. 7.9** Cause-of-death patterns of the four main global clusters compared to all-Russia. *Source* Vallin et al. (2005)

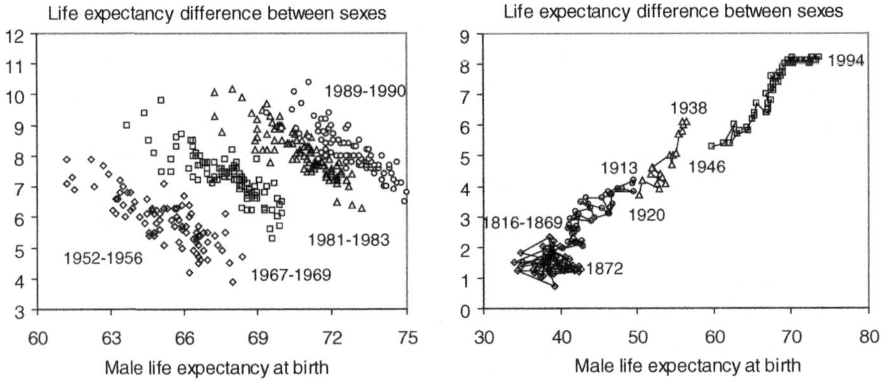


**Fig. 7.10** Relationship between the gap in male and female life expectancy and the level of male life expectancy as observed in France among its "Départements" at various censuses and by year from the beginning of 19th century to the end of the 20th. *Source* Vallin (2002)

1990), when studied through geographical analysis, it seems that the higher the level of life expectancy, the narrower the gap between sexes, as shown in the left-hand graph of Fig. 7.10, at four successive French censuses. Each time, the cloud of points is clearly oriented in that direction. But from one census to the next the cloud goes up in the reverse direction, exactly like the sex gap for life expectancy in France, shown in the right-hand graph of Fig. 7.10: when life expectancy increases, the sex gap also increases.

This strange paradox gets its explanation when looking at causes of death as shown through a comparison between France and Russia (Meslé and Vallin 1998). As in many developed countries, the historical rise of life expectancy in France has been mainly explained by the decrease in infectious diseases up to the 1960s and then in the cardiovascular diseases. Sex differences increased in parallel, firstly because of the increase in male bad habits (excessive alcohol consumption, smoking, risky driving, etc.) and specific occupational exposure but also, more recently, because women were able to benefit from the cardiovascular revolution sooner than males. Altogether, it was quite logical for the sex gap to grow at the same time as life expectancy, at least until very recently. By contrast, French geographic variations in mortality are strongly related to the negative effects of alcohol consumption, which is also an important explanation for the sex gap. Quite logically, the larger the gap, the higher the level of mortality and the lower the life expectancy. It is for a very similar reason that in recent decades the Russian sex gap increased sharply while the level of life expectancy decreased, both trends being largely driven by alcohol consumption.

Sex differences are also a field where the best experts in demography have made wrong predictions. In the 1960s, Coale and Demeny (1966) published model life tables where after a slight increase in sex differences from lower to higher levels of life expectancy the relation was reversed at the higher levels (Fig. 7.11). They argued that since behaviours of both sexes in terms of smoking, driving, working, etc. are now converging, in the same way, life expectancy should also converge. Soon after they published their tables, trends observed in many developed countries were exactly the opposite, especially in countries like Sweden or the U.S.

In fact, the period with increasing sex differences in life expectancy relates to the greater negative effects of smoking for men, but also that of greater benefits from the cardiovascular revolution for women. And, actually, Coale and Demeny were not so

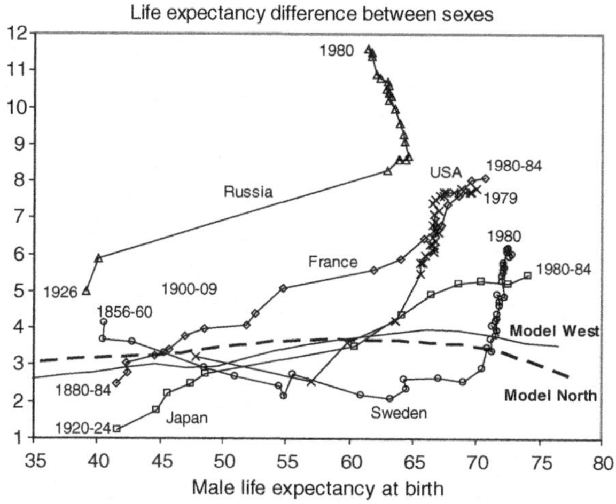


**Fig. 7.11** Relationship between the level of male life expectancy at birth and the difference between male and female life expectancy according to Coale and Demeny model life tables and three national historical experiences (up to the 1980s). *Source* Vallin (2002)

wrong, but only right too soon. As soon as the late 1960s in England and Wales, and more generally since the 1980s in other English-speaking countries and in European Nordic countries, male tobacco consumption started to fall while men increasingly copied healthy female behaviour in relation to cardiovascular diseases. Figure 7.12 shows the results in terms of sex gap/level of life expectancy trends.

## Social Groups

Inequality in life expectancy by social group is one of the most unacceptable ones. Common sense tells us that even if economic differences between individuals in our capitalist world are unavoidable or even, for some economists, quite necessary, at least inequality in death should be eliminated. Indeed most developed countries have implemented policies to reduce them, not only by encouraging private insurance that aims to socialise the risks of having to pay for health, but also by

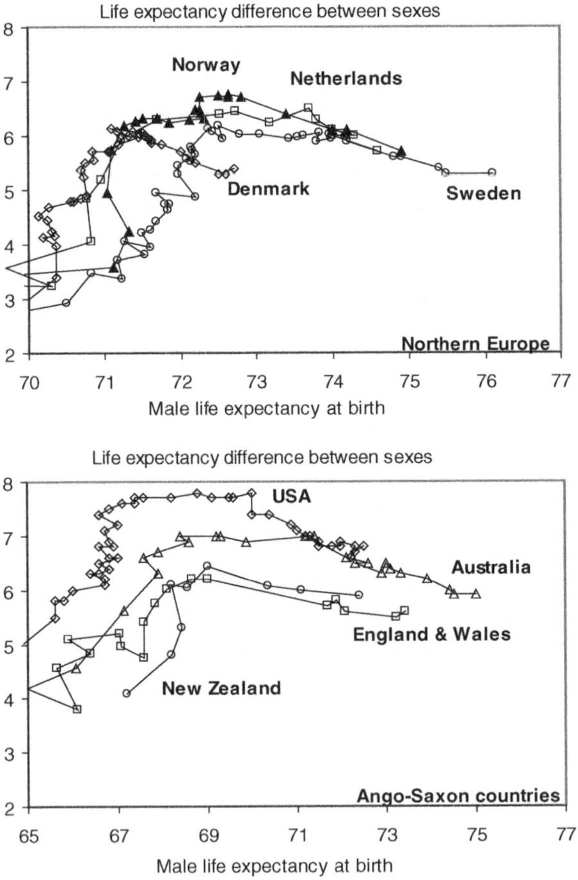


**Fig. 7.12** Relationship between the level of life expectancy at birth and the difference between male and female life expectancy in some English-speaking and Northern European countries in the three last decades of the twentieth century. *Source* Vallin (2002)

setting up compulsory systems of social security, one of the masterpieces of the European "welfare state". Especially, after Pasteur's revolution, as medical science became more and more efficient in fighting infectious diseases, the leading cause of death at that time, the idea that it was possible to extend the full benefits of new medical techniques to the population as a whole became more and more popular. New policies were introduced aiming to reduce dramatically social inequality in health and life expectancy. Especially after World War II, not only in the Communist region but also in the West, all European countries established social health systems which, thanks to the spectacular success of antibiotics, offered great promise of future progress.

Half a century later, the hope of eliminating or even of gradually reducing social differences in life expectancy had vanished completely in the face of results like those illustrated in Fig. 7.13 for France, which compares recent trends in male life expectancy at age 35 among 7 broad socio-occupational groups to that of France as a whole. Unfortunately, only 3 points in time are available for social groups (1980, 1988 and 1996) but that is enough to be very pessimistic. Indeed, while the indicator increased dramatically at the national level (from about 35 years in 1955 to 43 in 2005) it also increased for each of the social categories, but the pace of increase was not faster for the lowest categories than for the highest. On the contrary, the social divide seems greater than ever, since life expectancy increased more rapidly for the highest groups (senior management and higher intellectual occupations) than for the lower ones. Consequently, while, in 1996, male life expectancy at age 35 of the lowest group (inactive people) was at the same level as the national one in 1810 (almost two hundred years before!), it is at the national level of 1980 for "manual workers", and 1990 for "clerical workers". Conversely, in the same year, 1996, the 2005 national level had been already reached by "self-employed" and "intermediate occupations", and although we might expect the national level to reach the 1996 level of "agricultural occupations" quite soon, it is much less sure that the national level will reach the one observed for the highest occupational category in 1996 within a few decades.

## 7.3 What is the Explanation?

How can we explain such large differences in life expectancy among and within European countries that are, along with Japan and overseas English-speaking countries, the most advanced ones in terms of health and welfare systems, after more than one or two hundred years of tremendous medical, economic and social progress? This in itself is a big challenge. Is it possible to explain everything together? Certainly not, but let us try.

And, perhaps, let us do even more. Although this paper is devoted to Europe, it would be surprising if the case of Europe followed different rules than the rest of the world. Fortunately, Europe was the first to radically change the past traditional relationship that human beings had with their own health and length of life. For

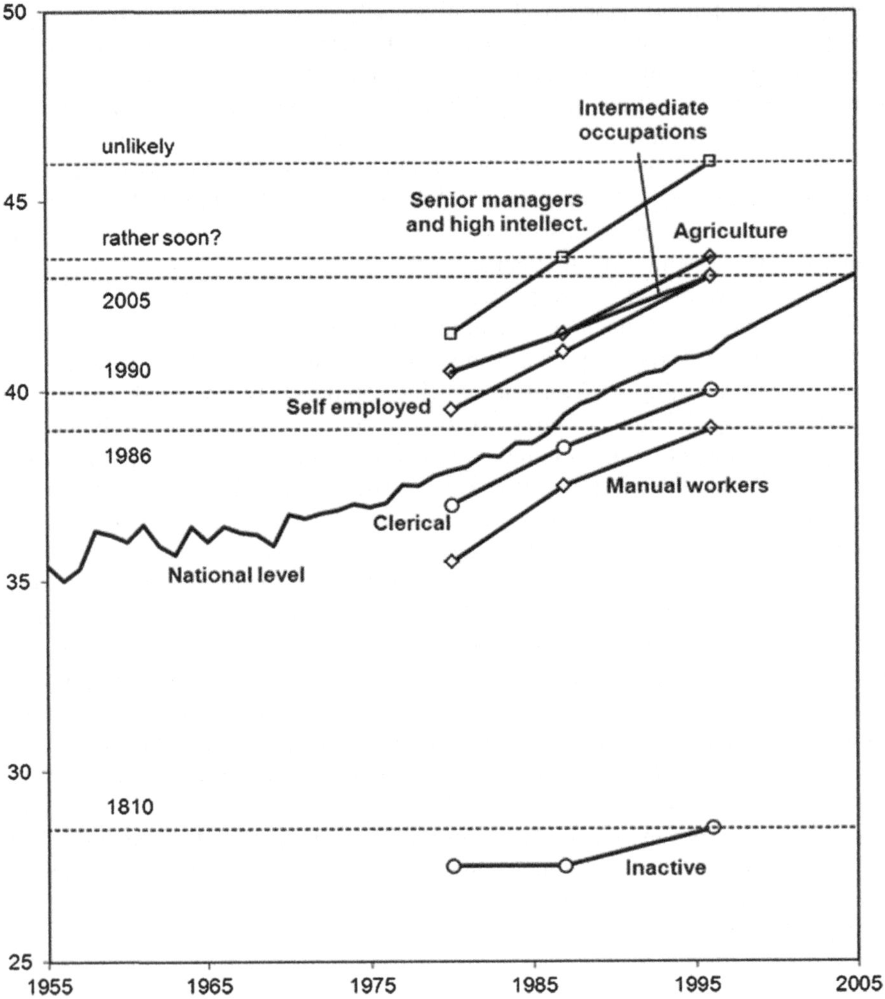


**Fig. 7.13** Male life expectancy at age 35 by occupational groups, as compared to the national trend in France (1955–2005)

that reason, Europe is the main "laboratory" to test theories but we should not forget that any general explanation must take into account the whole story, including that of the developing world.

A basic idea can help us: *Different histories have resulted in the current diversity*. If that is true, then we can look at general theories aiming to explain historical trends, starting with Omran's (1971; 1983) epidemiologic transition theory but also looking at the attempts to correct or to replace that theory after it failed to explain more recent developments, and finally summarising the more pragmatic ideas recently proposed by France Meslé and myself.

### *7.3.1 From Epidemiologic Transition to Health Transition*

Omran (1971) was the first to explicitly propose a general paradigm to explain the historical mortality decline that was the first pillar of the demographic transition, opening the door to the need for fertility control. According to his so-called *epidemiologic transition theory* (Omran 1971), in the process of the industrial revolution and of socio-cultural modernization, any society moves from an initial "age" to a final one by crossing an intermediate stage where the epidemiologic transition takes place. Thus, all countries have experienced or will experience soon:

- First, the age of *pestilence and famine* characterized by a very high current mortality, often exacerbated by severe fluctuations due to epidemics and food shortages; under these conditions, long-term mean life expectancy barely reaches 30 years;
- Second, the age of *pandemics receding* when modern medicine vanquishes most infectious diseases, especially at young ages, resulting in a dramatic fall in infant and child mortality and in a doubling of life expectancy from around 30 to about 60 years;
- Third, the age of *degenerative diseases and man-made diseases* where the life expectancy increase slows down because the returns of continuing decline in infectious mortality become smaller and because modern societies produce "man-made diseases" like alcoholism, smoking, traffic accidents, etc. that develop very rapidly and tend to reduce life expectancy that very rarely exceeds 70 or 75 years.

In Omran's view, this "end of the transition" was a strong reason to think that in a relatively short time, step by step, all the countries that have not yet approached such a probable limit for life expectancy will converge towards the maximum almost reached already by the most advanced ones.

Obviously, facts decided differently. On the one hand, as soon as Omran published his founding article, new progress in the field of the so-called "degenerative diseases" opened the door to a new life expectancy increase, while, step by step, modern societies proved capable of controlling man-made-disease epidemics. But on the other hand, in Europe, as we saw here above, Central and Eastern European Communist countries failed to exploit that new opportunity and were unable to interrupt the rise in man-made diseases. Even more striking, two decades later, the quasi eradication of infectious diseases promised by Omran was very abruptly challenged by the AIDS epidemics, especially in Africa.

Was it enough to expand Omran's theory by adding new ages to include all these unexpected events in its explanatory framework? Olshansky and Ault (1986), for example, proposed to add the *age of delayed degenerative diseases* to take into account the cardiovascular revolution. To do so, almost at the same time, Rogers and Hackenberg (1987) called that step *the hybristic stage* to further highlight the role of diseases related to behaviour and lifestyle. Later on, Olshansky et al. (1998) even suggested a "fifth age", that of *re-emergence of infectious diseases* to take AIDS onboard, while Omran himself (1998) described a new *age of aspired quality of life with paradoxical longevity and persistent inequities*. Finally, in the same article, Omran added a "sixth age", *towards equity and quality*, simply to fit with the WHO *Health for all* catchword. For sure, in the absence of common background making it possible to join up all the pieces of the puzzle, it was better to rethink the whole perspective, as other authors did, mainly in the 1990s.

One important change, proposed by Frenk et al. (1991) in the first issue of the new journal *Health Transition* founded by Jack Caldwell, was to replace the word *epidemiologic* by *health*, which is not a simple matter of semantics but aims to emphasize that the rise in life expectancy is not only related to a change in the distribution of pathologies due to epidemiological events but also to changes in social and behavioural factors and in the economic and political capacity of societies to react to epidemiological facts. The authors draw upon the health transition concept proposed 20 years before by Lerner (1973) to widen that of epidemiologic transition and propose to combine different levels of analysis: "systemic", "societal" "institutional" and "individual". No doubt that this point of view is crucial to understand the different and sometimes contradictory trends in life expectancy. However, the authors did not make any attempt to demonstrate their idea on the basis of empirical data.

A second interesting idea came from Horiuchi (1999) who suggests considering several successive epidemiologic transitions that modified the pathological structure of the population at the same time as society moved forward from one technological era to a new one. The paradigm is based on the idea that each great technological era, namely "hunting and gathering", "agriculture", "industry", "high technology", and "future", is associated with a specific type of society and dominated by a major cause of death, respectively *violence*, *infection*, *cardiovascular diseases*, *cancers*, *and aging*. It is interesting to note that Horiuchi's idea is rather close to the concept of pathocenosis proposed by the historian of medicine Grmek (1969), who considered periods of equilibrium between pathogenic agents and short transitions between one pathocenosis and the next. But the big difference is that modern epidemiologic or health transitions lead from a lower to a higher level of life expectancy, while the change from one historical pathocenosis to the next leaves the high level of mortality quite unchanged.

### 7.3.2 Divergence–Convergence

In a study co-authored by France Meslé and myself, we described a possible general explanation for the great diversity of levels and trends in life expectancy by giving greater importance to empirical data. It seems to us that the latter suggest a succession of divergence-convergence movements that take place at different times from population to population (Vallin and Meslé 2004; 2005).

Our research is based on two initial ideas:

1. In the face of current heterogeneity, many studies have discussed the hope that a process of convergence towards the best life expectancy will occur. In our view, before speculating on convergence, it is important, first, to recognize that heterogeneity itself can result from a previous divergence process, and, second, to imagine what the cause of this divergence was.
2. Any major factor of improvement in life expectancy results in a phase of divergence, since some countries (or populations) are more ready to benefit from it than the others. However, after some time, those that lagged behind catch up with the pioneers in a convergence phase.

Let us look at history in the light of such ideas. At least two, but very likely three major movements of divergence-convergence quite obviously occurred or started within the two last centuries of human history. The first one corresponds quite well to what Omran called the epidemiologic transition: the control of infectious diseases. The second one is the cardiovascular revolution. The third could be the fight against old-age mortality. None of these three processes involved all the different countries (or populations) at the same time and nothing strictly requires that one process be completed before another one starts.

## First Process: The Control of Infectious Diseases

Figure 7.14 displays life expectancy trends of a series of industrialized countries from the mid-eighteenth century to the mid-1960s. Not all the countries are represented to keep the drawing legible but this does not affect the general shape

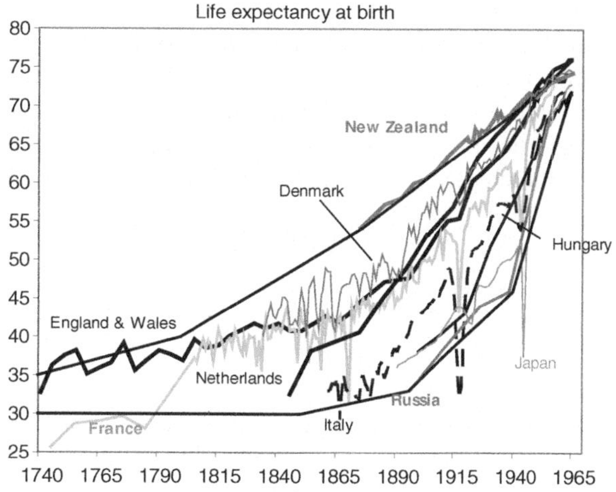


**Fig. 7.14** Long-term trends in female life expectancy for selected industrialized countries until the mid-1960s. *Source* Vallin and Meslé (2004)

suggested. A strong convergence of trajectories appears clearly from the years 1880s to the mid-1960s, between New Zealand at the top and Russia and Japan at the bottom, with all other countries rather well spread in-between.

It is less easy to see the divergence that occurred before the convergence because of the lack of complete historical series but the two cases represented here, France and England and Wales, very probably, over-estimate the maximum and under-estimate the minimum prevailing in the second half of the eighteenth century.[6] The interval between the two lines of Fig. 7.14 which were drawn to represent the upper and lower limits in between which the improvement of life expectancy in industrialized countries occurred thus gives a good idea of the complete process of divergence-convergence. A long discourse is not necessary to understand that such a movement is related to the industrialized countries' successful fight against infectious diseases. The basic scheme is rather well known: it is a continuum of changes in the fields of economics (from the agricultural progress of the mid-eighteenth century up to the industrial revolution), of medical technologies (from Jenner's vaccine to Pasteur's discoveries and to immunization, sulfamides and finally, antibiotics), and of social policies (from the generalization of education to the post-WWII social security systems). More interesting to note is that the timing of different countries' entry into that first movement of the health transition is linked to the timing of their modernization. England and Wales were the first, soon followed by France, the Nordic countries (here Denmark) and the Netherlands. Then the Mediterranean or Central European countries like Italy and Hungary started to catch up with the pioneers and finally Russia and Japan succeeded in doing the same, somewhat later but also much faster.

Without saying too much about developing countries that are not the focus of this paper, two things are important to underline. On the one hand, although they started to modernize later than the latest European countries, most of them are now (or will be very soon) catching up with the levels of life expectancy reached by the pioneers. They thus confirm that as soon as their economic and political situation enables them to benefit from rather simple and cheap technologies like immunization and antibiotics, they can also reduce infectious mortality dramatically. Many African countries, on the other hand, were so badly hit by the world economic crisis of the 1980s that they were unable to provide themselves with the basic conditions for doing so. Furthermore, those who were strongly affected by the AIDS epidemics experienced even more social and economic disruption and were totally unable to face the challenge. Consequently, life expectancy fell by 10, 15 or 20 years!

---

[6] Both of them come from historical demography reconstruction but the English methods tended to underestimate the role of migration while the French one could have over-estimated infant mortality.

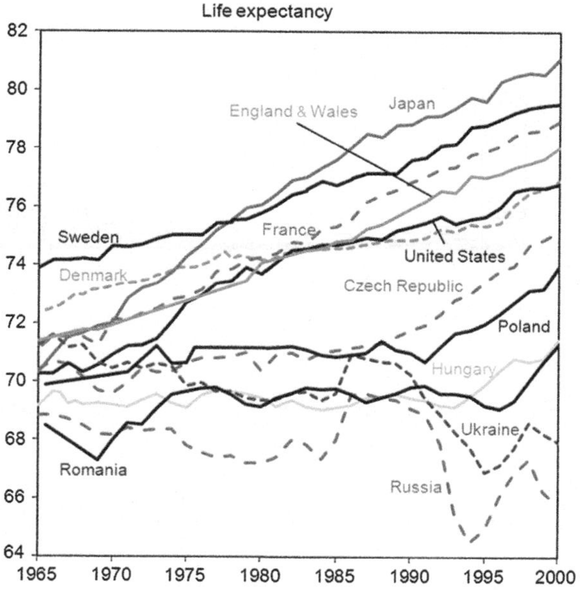


**Fig. 7.15** Trends in life expectancy (both sexes) in industrialised countries since 1965. *Source* Vallin and Meslé (2004)

## Second Process: The Cardiovascular Revolution

Contrary to the control of infectious diseases, the diverging trends caused by the cardiovascular revolution are striking (Fig. 7.15) while the converging ones are less so because the whole process is far from its end, even when the observation is limited to the industrialised world. Two main factors are responsible for the cardiovascular revolution: implementation of new technologies and behavioural changes. From a technological point of view, things are less simple than for infectious diseases. Here, progress can be attributed more to the rapid accumulation of a series of medical discoveries or improvements, from complex drug protocols to refined diagnosis techniques, advanced surgical procedures and sophisticated alert networks for rapid intervention. Active participation of the patients themselves is more important than for the fight against infection. But it is even more important for the necessary behaviour changes in terms of diet, exercise or medical screening for prevention, not to mention smoking and/or alcohol consumption.

In a first step, from the mid-1960s to the mid-1980s, the cardiovascular revolution almost immediately resulted in a strong divergence of life expectancy trends between a very steady increase in Western European countries (as well as in Japan and the U.S. of course) and a stagnation in the Central European countries under Communist regimes and even a deterioration in the Soviet Union (Fig. 7.15). In a second step, from the late 1980s or early 1990s, life expectancy started rising again in Central Europe. The pace observed in Poland and the Czech Republic suggests that these two countries could catch up with the U.S. or Denmark very soon and perhaps one day will reach the level of the most advanced pioneers (but this is probably another story as we shall see below).

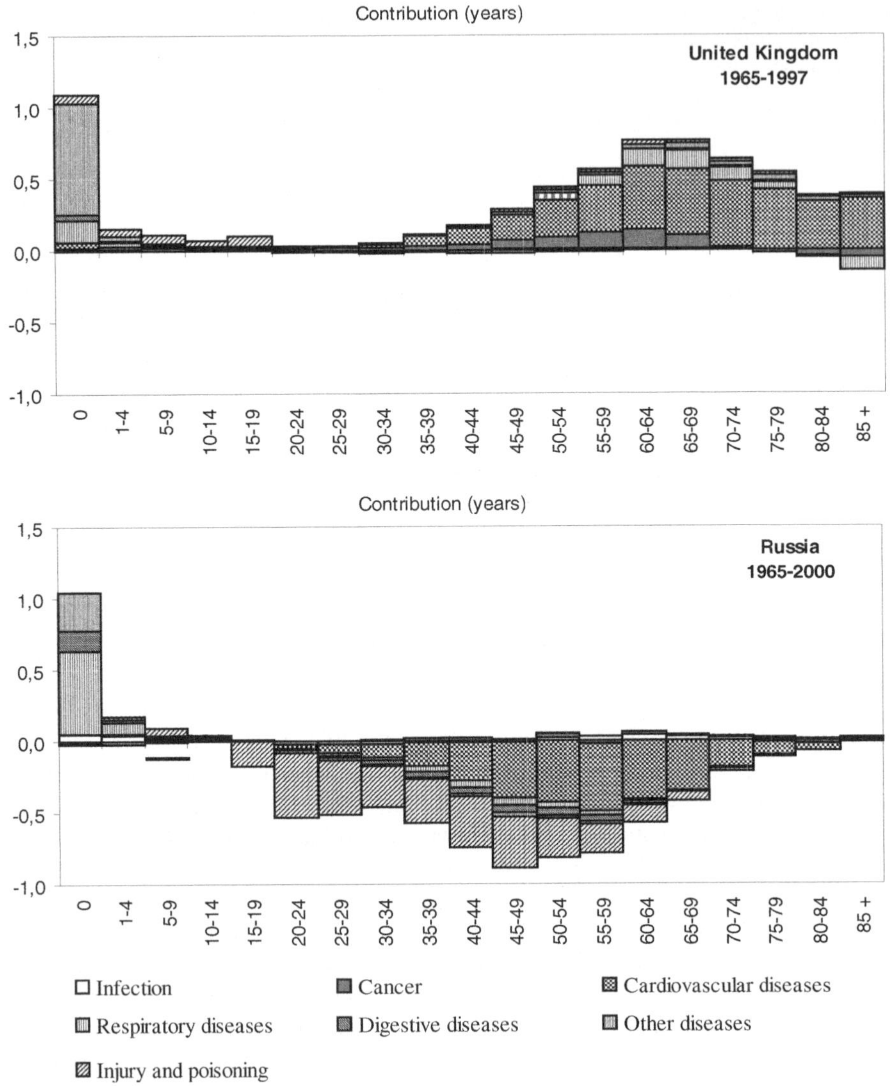


**Fig. 7.16** Contribution of age-and-cause-specific mortality in male life expectancy changes since 1965, in the United Kingdom and Russia. *Source* Vallin and Meslé (2004)

The cardiovascular revolution is not the only cause of the divergence between East and West. Figure 7.16 shows the impact of age-and-cause-specific mortality changes on life expectancy changes since 1965 in the United Kingdom and Russia. The role played by cardiovascular mortality at adult ages is massive both for the UK life expectancy increase and for the Russian life expectancy decrease. But the latter is also due to the increase in external causes of death (including alcohol abuse and violence) at younger adult ages. Besides that, both countries made life expectancy gains from a steady and substantial decrease in infant mortality that reinforces the life expectancy rise in the UK and moderates its decline in Russia.

Finally, Russia, and until the end of the 1980s, all Central and Eastern European countries, followed Omran's theory quite well: they had reached the "third age" where some decrease in infectious diseases is still possible but a substantial rise in life expectancy very improbable because of the pandemics of degenerative diseases and the spread of man-made diseases.

First, the former Communist countries were unable to benefit from the cardiovascular revolution for at least two reasons. Pushed by the Cold War to invest heavily in the military effort, they cut other budgets, including that of the health system, and were unable to provide access to new technologies to the whole population. At the same time, the necessary changes in individual behaviours were not really encouraged by a political context where people expected to receive everything from a central power that discourages any individual initiative. Second, these countries were also unable to stop the rise of man-made diseases, which often originated in habits that were more encouraged than discouraged by the political authorities.

It is more difficult to explain the reason why the life expectancy increase resumed in some of them (like Poland and the Czech Republic) but not in others (like Russia and Ukraine). The fact that the latter stayed longer under a Communist regime is not a sufficient explanation. More important is certainly the way these countries changed to a market economy. On the one side, Poland and the Czech Republic, had already attempted to change their economic and social system (with more or less success) well before the fall of the Berlin Wall and were better prepared for such a change, but they also transformed their economic and political system with less brutality and rapidly adopted positive institutional changes in the health system, whereas the Russian social and health services simply collapsed. Russia's transition towards a new and more efficient system will take more time.

## Third Process? The Fight Against Old-Age Mortality

Although the cardiovascular revolution is not yet completed, it is quite possible to think of a third divergence-convergence movement when looking at female life expectancy in the most advanced countries. Women are much more advanced than men: Fig. 7.17 shows that in 2000, male life expectancy in the most advanced countries was not yet at the level already reached by women in 1980. If some new change were to appear, it would very likely be among females.

Indeed, the right-hand graph of Fig. 7.17 suggests that a new divergence process has started among women, between countries like Japan, France or Italy, where life expectancy increases steadily throughout the period, and some other countries like the U.S., Denmark, and the Netherlands, where it is slowing down.

Actually, recent life expectancy increases are largely driven by the mortality decline at old ages. Let us see what happened in life expectancy at age 65 in four of these countries, Japan, France, U.S. and the Netherlands. Figure 7.18 is quite

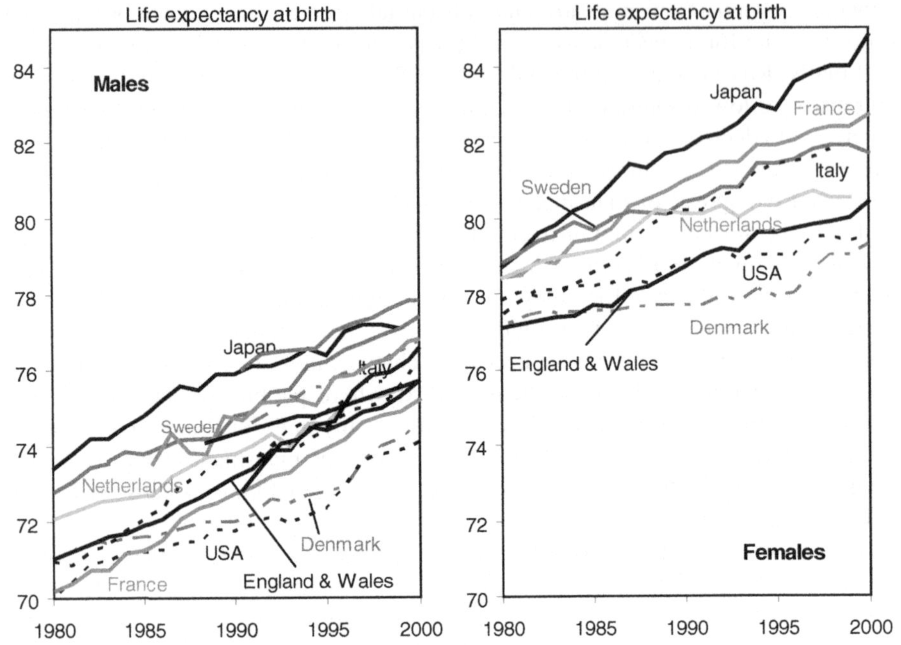


**Fig. 7.17** Male and female life expectancy at birth in 8 industrialized countries, 1980–2000. *Source* Vallin and Meslé (2004)

**Fig. 7.18** Female life expectancy at age 65 in 4 industrialized countries, 1980–2005. *Source* Meslé and Vallin (2006)

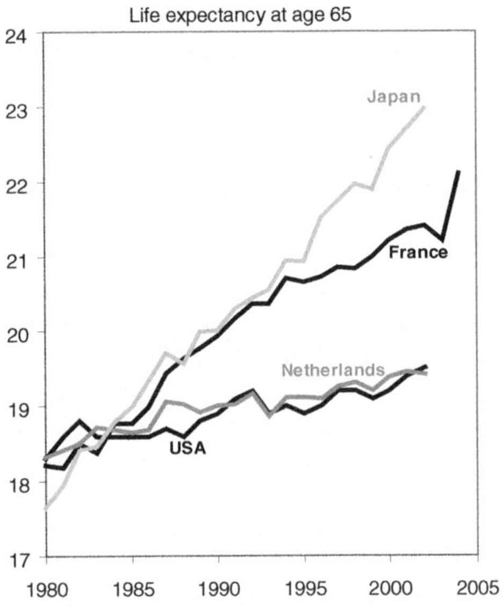

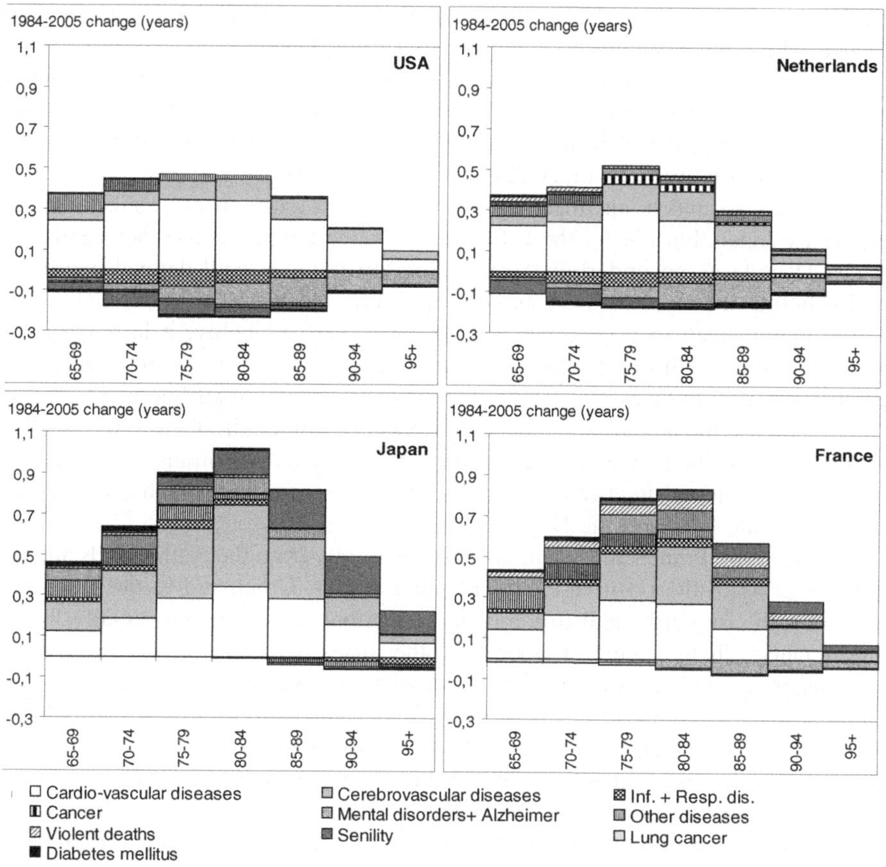


**Fig. 7.19** Age and cause components of life expectancy changes 1984–2005: increasing weight of mental disorders in the U.S. and the Netherlands. *Source* Glei et al. (2009); Meslé and Vallin (2006)

impressive: Japan and France on the one side, the U.S. and the Netherlands on the other, started to diverge dramatically from the early 1980s onwards.

Figure 7.19 displays the respective impact on life expectancy at age 65 of cause-and-age-specific mortality changes over the 1984–2005 period in each of the four countries. Gains in France and especially Japan are very impressive when compared to both, the U.S. and the Netherlands, where smaller gains are offset by sizeable losses. But the differential impact of causes by age is also remarkable.

First of all, whereas in France and Japan, very few age groups exhibit the negative effect of some causes, the impact of which is also very small, the U.S. and the Netherlands are hit by mortality increases for several causes across all age groups which accumulate into substantial losses. The greatest negative impact is due to mental disorders (including Alzheimer's disease) while infectious and respiratory diseases[7] are the second source of losses, closely followed by lung cancer, which has virtually no effect in France and Japan.

The second difference between the U.S. and the Netherlands on the one hand, and France and Japan on the other, is that for the former two countries most gains are due to the decline of heart diseases, while France and Japan enjoyed greater declines in mortality from other diseases of the circulatory system.

Senility also makes an important contribution to gains in $e_{65}$ for France and especially Japan, but not for the U.S. and the Netherlands because they rarely use this ICD code as an underlying cause. Senility is not a well-defined cause but redistributing it proportionately as generally done for the other ill-defined causes would mainly lead to a false perspective since, very probably, it is much more related to mental disorders than any other cause (Meslé 2006) or, perhaps, to cerebrovascular diseases. No doubt that an appropriate redistribution of senility would increase the already huge positive impact of "other circulatory diseases" in France and Japan, create a positive impact for mental disorders in Japan and change its negative effect into a positive one in France, while nothing would be changed in the U.S. and the Netherlands.

Furthermore, France and Japan differ substantially from the Netherlands and the U.S. in terms of the positive effects of other causes. Compared to the latter two countries, the former benefit from greater reductions in diabetes (especially Japan) and in causes that are here grouped as "other diseases".

Thus, a new process of divergence seems to have started before the end of the cardiovascular revolution. Gains in the U.S. or Dutch life expectancy are almost exclusively due to cardiovascular mortality decline (especially heart diseases) and they are significantly eroded by losses due mainly to mental disorders. On the contrary, almost no losses affect France and Japan, where gains on cardiovascular diseases are more on cerebrovascular than heart conditions and much more concentrated in the very old ages, and where many other causes of death are sources of gains. Especially if French or Japanese senility can be largely assimilated to mental disorders, one could say that the United States and the Netherlands entered the rising pandemic of mental disorders foreseen by Kramer (1980) almost as soon as he published his article, while France and Japan had already found the way to escape it. More generally, a societal change could be responsible for this new process of divergence: the healthcare of the oldest people has improved substantially in France and Japan during the last two or three decades, perhaps not so much in the U.S. and the Netherlands. It has become less and less acceptable to let old people die without medical intervention. Such change has a technical basis, since medical intervention has become less and less aggressive. But it has also a cultural basis and probably some countries are more ready than others not only to use all medical technologies to gain additional years of life among old people, but also to pay enough attention to them in their daily life and to small health problems to keep them in good shape for longer. In France, the 2003 heat wave resulted in a

---

[7] In fact, respiratory diseases account for most of the observed divergence.

paradoxical benefit: more lives were saved the year after than lost in 2003, just because the crisis obliged health services and society as a whole to see things differently.

## What About Sub-National Differences?

If the complexity of mortality trends becomes somewhat more understandable when considering these divergence-convergence processes caused by major advances in health, can such processes be helpful to interpret within-country mortality trends and differentials as well? In-depth exploration of this question is beyond the scope of this paper, but we will look briefly at selected examples of internal geographical variations, gender differentials, and social inequalities.

### Geography

It is a matter of record that in the case of French *départements*, substantial convergence is observed across the nineteenth century (Caselli and Vallin 2006; Vallin 1990). In Fig. 7.20, female life expectancy trends by French *départements* are compared to the upper and lower limits of national trends in industrialised countries already plotted for Fig. 7.14.

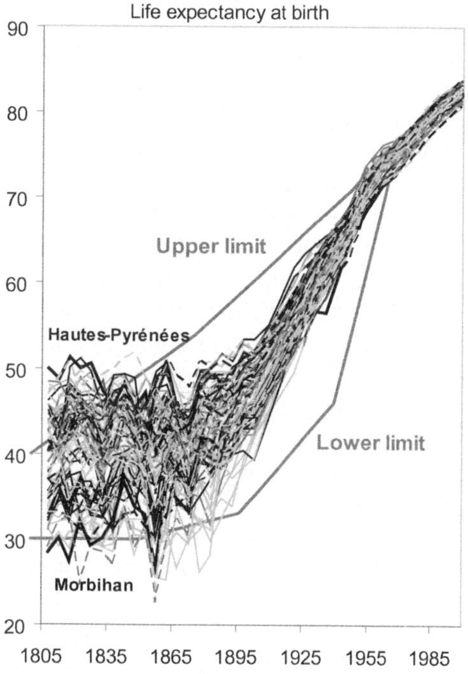


**Fig. 7.20** Trends in female life expectancy by French *départements* compared to the upper and lower limits of national trends in industrialised countries. Caselli and Vallin (2006)

At the beginning of the nineteenth century, notwithstanding the low national level of life expectancy (less than 40 years), the difference between *départements*' maximum and minimum values was more than 20 years throughout the first half of the nineteenth century, with a quite stable standard deviation of approximately 5 years. Unlike what was being seen between countries, no divergence appears in the early nineteenth century among French *départements*. It may well be that a divergence stage occurred before 1800, since the likelihood that some French *départements* always enjoyed life expectancy above 50 years is remote. However, it is also highly likely that, in the past, mortality differentials between small geographical units were much wider than international variations. Specific local hazards, in terms not only of climatic variation but also of economic and social environment, may have played a very important role. For example, the quality of drinking water and practices in disposal of liquid waste varied widely from place to place, and had a material impact on life expectancy. To begin with, therefore, there was probably greater heterogeneity within each country than between countries.

In the second half of the nineteenth century, by contrast, convergence between French *départements* is even more dramatic and rapid than it was between countries. This is probably the consequence of centralized economic, social and health policies that were very successful in spreading most benefits of health advances. It was an era of generalization of free trade on the domestic market, rapid expansion of transport networks (roads, railways) and development of food preservation techniques that raised food safety to near-equal levels nationwide. Likewise, public policies were implemented to develop drinking water supplies, sewerage systems and public hygiene up and down the country. Finally, the health advances brought by Pasteur's revolution, including the development of personal hygiene, were rapidly disseminated nationwide through the new education policy initiated by Jules Ferry, bringing in compulsory, free schooling.

In the second half of the twentieth century, and especially after the end of the 1960s, it might have been expected that the cardiovascular revolution would first result in a new divergence, as it had at the international level. Far from increasing, standard deviation for female life expectancy continues to decrease slightly from 1.1 in the 1950s to 0.7 in the 1990s. Internal geographical variations appear to have been relatively impervious to the dramatic change in French cardiovascular mortality. At that time, the major determinants of that mortality decrease were quite evenly distributed country-wide, thereby precluding any divergence. New health behaviours (diet, physical exercise, reducing alcohol consumption, etc.) spread across the country. Modern preventive medicine and treatments were well distributed among *départements*.

Indeed, an important difference appears here between national and local levels. At the national level, diverging processes are caused by the diversity of social, cultural, economic and political capacities to initiate new technologies and practices or to benefit from them. In a centralized state such as France, sub-national entities do not (or at least no longer) have enough political or cultural autonomy to react to health innovation at significantly different paces. It would be interesting to look at a federal state, like the U.S.

## Gender Differentials

The left-hand side of Fig. 7.21 shows trends in absolute differences between male and female life expectancy in three Nordic countries and the Netherlands. The Swedish series are the longest, but the other three display very similar trends since the mid-nineteenth century. In Sweden, gender differentials first increased in the very early nineteenth century, decreasing thereafter to stabilise from the end of the nineteenth century up to World War II. Then came a fresh steady rise from the 1950s to the 1970s, followed by a renewed drop in the 1980s and 1990s. Can these wide fluctuations be interpreted in the same way as for the countries discussed above? It is clear that while improving their social status, women benefited increasingly from advances in the reduction of infectious diseases, especially at younger ages, but it is not so obvious to interpret men lagging and then catching up with women in terms of infectious diseases. The next wave underlines that delayed adoption of not only favourable but also unfavourable changes can contribute to a divergence-convergence process: from the 1950s to the 1980s, men lagged first because they were sooner and more massively affected by the rise in man-made diseases and second because women were more ready to benefit from the means to fight cardiovascular diseases. Trends in differentials dropped again in the early 1980s when man-made diseases came under control and men progressively adopted the new behaviours favourable to the reduction of cardiovascular diseases.

The same is true of Norway, Denmark, and the Netherlands, at least for the period covered by available data. On the right-hand part of Fig. 7.21, very similar trends appear for Anglo-Saxon countries (England and Wales, Australia, New Zealand, Canada and the U.S.) for the more recent wave of changes.

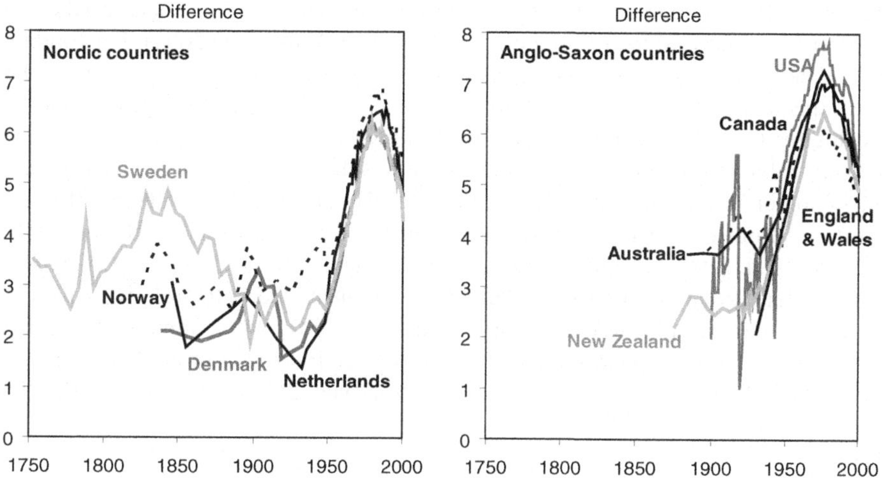


**Fig. 7.21** Trends in the gap between male and female life expectancy in selected countries. *Source* Vallin and Meslé (2005)

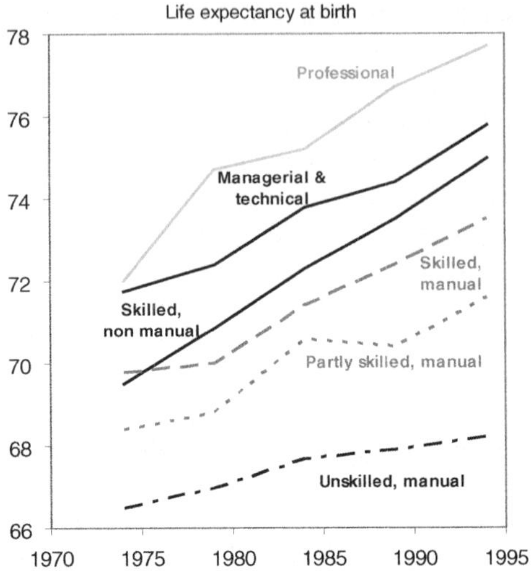


**Fig. 7.22** Male life expectancy at birth in England and Wales by socioeconomic category during the period 1972–1996. *Source* Valkonen (2001)

## Social Classes

Although it would be much more complicated to attempt the same type of interpretation for socio-economic differentials, there is good reason to believe that life expectancy differences between social classes, levels of education and so on, could also have varied over time according to the ability to benefit from the different stages of the health transition.

Much of the difficulty here stems from the absence of long historical datasets and also from the poor comparability of existing data at different times.

However, for recent decades, especially the post-WWII period where data do exist for different countries, most studies show an increase in the social gradient of mortality, as assessed for Europe in an overview by Valkonen (2001). For England and Wales, for instance, regular series of social mortality surveys show clearly that life expectancy at birth increased for all social categories, but much more so for the higher than for the lower ones, resulting in a widening gap (Fig. 7.22). As Tapani Valkonen pointed out, "mortality from ischemic heart disease (IHD) was a major factor in the increase in mortality differentials among men aged 20 to 64. Mortality from IHD decreased by more than 50 % in Classes I and II from 1970/1972 to 1991/1993, but only 33 % in Class IV and 3 % in Class V.[8] Changes in lung cancer, stroke, accidents, and suicide also contributed to the relative widening of the socio-economic mortality gap" (Valkonen 2001, 245). Some of the socio-economic differences in mortality are due to selection of socio-economic groups by individual health status (Fox et al. 1985; Valkonen 2001; Vallin 1979). But that

[8] English studies distinguish 6 social classes : professional (I), managerial and technical (II), skilled non-manual (IIIN), skilled manual (IIIM), partly skilled manual (IV), and unskilled manual (V).

does not explain the widening of the gap. It is highly likely that the higher socio-economic categories were more able to take advantage of the cardiovascular revolution (either by adapting their behaviours or accessing the most sophisticated new medical technologies of screening and treatment), and to avoid man-made diseases as well.

## 7.4 Conclusion: A Rather Pessimistic View

After identifying three major divergence-convergence processes (corresponding respectively to the vanquishing of infectious diseases, the cardiovascular revolution, and finally the fight against old-age mortality) as three important features of historical life-expectancy trends, it becomes somewhat easier to understand the extreme diversity of levels, trends and patterns of mortality observed between and within countries. Both the current diversity and its sometimes paradoxical appearance are easier to understand, especially when considering that a new divergence-convergence process can start even if the previous one has not ended. This is true among entities (countries or sub-populations): the cardiovascular revolution started before most developing countries completed infectious disease control. But it is also true for a specific entity. For example, the cardiovascular revolution had not yet ended either in France or in Japan when these countries started to diverge from the U.S. or the Netherlands in terms of old-age mortality. Furthermore, one specific process can be interrupted if there is a reversal in the social, economic or political factors that allowed it to start. For example, African progress in the control of infection was held back by the 1980s world economic crisis and then severely eroded by the AIDS epidemics. It is no longer surprising that current inequalities in life expectancy are greater than ever.

What about the future? The three convergence-divergence processes identified here are not, by any means, an exhaustive list of all possible processes of this kind. It has been especially surprising to observe the beginning of a third transition in the field of mental diseases, when we might have expected the next challenge to come from cancers, which became the leading cause of death after the decline of cardiovascular diseases. No doubt that in the future, a more successful fight against cancer will generate a new divergence-convergence wave. And we can think of many other old or new pathologies as the next potential candidates. Furthermore, a completely new type of divergence-convergence could occur if the dream of lengthening the maximum human lifespan ever became a reality. No doubt that each of these possible innovations will open new doors to new divergences in favour of a few happy winners. Unless the whole process of health improvement finally reaches an impassable biological limit, we are likely to witness in the future new inequalities in health and life expectancy among countries and possibly within countries.

The only encouraging finding is that internal geographical differences can be reduced dramatically towards relatively sustainable homogeneity as the consequence of unified policies. But, even if international differences could be reduced in the near future at the level of the European Union, or, much later, at a global level by a world government, it would be hardly the case for differences among socioeconomic categories and across other criteria that support the fundamentally non-egalitarian economic model that prevails today.

## References


Caselli, G., & Vallin, J. (2006). Geographic variations of mortality. In G. Caselli, J. Vallin, & G. Wunsch (Eds.), *Demography: Analysis and synthesis: A treatise in population studies* (Vol. 2, pp. 207–234). New York: Academic.

Coale, A. J., & Demeny, P. (1966). *Regional model life tables and stable populations*. New Jersey: Princeton University Press.

Council of Europe. (2006). *Recent demographic developments in Europe, 2005*. Strasbourg: Council of Europe.

Duthé, G., Badurashvili, I., Kuyumjyan, K., Meslé, F., & Vallin, J. (2009). An attempt to re-estimate recent mortality trends in two Caucasian countries. *Demographic Research, 22*(23), 691–732.

Fox, A. J., Goldblatt, P. O., & Jones, D. R. (1985). Social class mortality differentials: Artefact, selection or life circumstances. *Journal of Epidemiology and Community Health, 39*(1), 1–8.

Frenk, J., Bobadilla, J. L., Stern, C., Frejka, T., & Lozano, R. (1991). Elements for a theory of the health transition. *Health Transition Review, 1*(1), 21–38.

Glei, D., Meslé, F., & Vallin, J. (2009). Diverging trends in life expectancy at age 50: A look at causes of death. In E. M. Crimmins, S. H. Preston, & B. Cohen, (Eds.), *International differences in mortality at older ages: Dimensions and sources*. Panel on understanding divergent trends in longevity in high-income countries. National Research Council, Committee on Population, Division of Behavioral and Social Sciences and Education. (pp. 17–67) Washington, DC: The National Academies Press.

Grmek, M. D. (1969). Préliminaires d'une étude historique des maladies. *Annales ESC, 24*(6), 1473–1483.

Horiuchi, S. (1999). Epidemiological transitions in human history. In J. Chamie & R. L. Cliquet (Eds.), *Health and mortality: Issues of global concern* (pp. 54–71). New York: United Nations.

Jasilionis, D., Shkolnikov, V., Andreev, E., Jdanov, D. A., Ambrozaitiene, D., Stankuniene, V., et al. (2007). Socioeconomic mortality differentials in Lithuania: Results obtained by matching vital records with the 2001 Census data. *Population-E, 62*(4), 597–646.

Kramer, M. (1980). The rising pandemic of mental disorders and associated chronic diseases and disabilities. *Acta Psychiatria Scandinavica 62*(S285): 282–297.

Lerner, M. (1973). Modernization and health. A model of the health transition. Paper to the American Public Health Association Conference, San Francisco.

Meslé, F. (2006). Causes of death among the oldest-old: Validity and comparability. In J. Robine, E. M. Crimmins, S. Horiuchi, & Y. Zheng (Eds.), *Human longevity, individual life duration, and the growth of the oldest-old population*. International Studies in Population (Vol. 4, pp. 191–214). Dordrecht: Springer.

Meslé, F., & Vallin, J. (1998). Évolution et variations géographiques de la surmortalité masculine: Du paradoxe français à la logique russe. *Population-F, 53*(6), 1079–1102.

Meslé, F., & Vallin, J. (2002). Mortality in Europe: The divergence between East and West. *Population-E, 57*(1), 157–198.

Meslé, F., & Vallin, J. (2006). Diverging trends in female old-age mortality: The United States and the Netherlands versus France and Japan. *Population and Development Review, 32*(1), 123–145.

Oeppen, J., & Vaupel, J. W. (2002). Broken limits to life expectancy. *Science, 296*(10), 1029–1031.
Olshansky, S. J., & Ault, B. (1986). The fourth stage of the epidemiologic transition: The age of delayed degenerative diseases. *The Milbank Quarterly, 64*(3), 355–391.
Olshansky, S. J., Carnes, B. A., Rogers, R. G., & Smith, L. (1998). Emerging infectious diseases: The fifth stage of the epidemiologic transition? *World Health Statistics Quarterly, 51*(2/3/4): 207–217.
Omran, A. R. (1971). The epidemiologic transition: A theory of the epidemiology of population change. *The Milbank Memorial Fund Quarterly, 49*(4), 509–538.
Omran, A. R. (1983). The epidemiologic transition theory: A preliminary update. *Journal of Tropical Pediatrics, 29*(6), 305–316.
Omran, A. R. (1998). The epidemiologic transition theory revisited thirty years later. *World Health Statistics Quarterly, 51*(2/3/4): 99–119.
Rogers, R. G., & Hackenberg, R. (1987). Extending epidemiologic transition theory. *Social Biology, 34*(3/4), 234–243.
Valkonen, T. (2001). Trends in differential mortality in European countries. In J. Vallin, F. Meslé, & T. Valkonen (Eds.), *Trends in mortality and differential mortality* (pp. 185–321). Strasbourg: Council of Europe.
Vallin, J. (1979). Facteurs socio-économiques de la mortalité dans les pays développés. In World Health Organization (Ed.), *Délibération de la réunion sur les facteurs et les conséquences socio-économiques de la mortalité.* El Colegio de Mexico, Mexico, June 19–25 (pp. 266–302). New York: WHO.
Vallin, J. (1990). Quand les variations géographiques de la surmortalité masculine contredisent son évolution dans le temps. *Espace, Populations, Sociétés, 3*(17), 467–478.
Vallin, J. (2002). Mortalité, sexe et genre. In G. Caselli, G. Wunsch, & J. Vallin (Eds.), *Démographie: Analyse et synthèse, III, Les déterminants de la mortalité* (pp. 319–350). Paris: INED.
Vallin, J., & Meslé, F. (2004). Convergences and divergences in mortality. A new approach to health transition. *Demographic Research, 2*(2): 12–43 (Special collection).
Vallin, J., & Meslé, F. (2005). Convergences and divergences: An analytical framework of national and sub-national trends in life expectancy. *Genus, 61*(1), 83–124.
Vallin, J., & Meslé, F. (2008). Les plus faibles mortalités: Un prédicteur des progrès à venir? *Population, 63*(4), 647–682.
Vallin, J., & Meslé, F. (2009). The segmented trend line of highest life expectancies. *Population and Development Review, 35*(1), 159–187.
Vallin, J., Andreev, E. M., Meslé, F., & Shkolnikov, V. (2005). Geographical diversity of cause-of-death patterns and trends in Russia. *Demographic Research, 12*(13), 324–380.

# Chapter 8
# The Occupational Mobility of Return Migrants: Lessons from North America

David P. Lindstrom

## 8.1 Introduction

Migrant labor is widely considered by population experts and policy makers as essential for meeting the economic challenges posed by population aging in Europe, and in many European countries migration is already a major component of demographic change (Bengtsson and Scott 2011; Bijak et al. 2008; Coleman 2006; Coleman and Rowthorn 2011). Recognition of the critical role of migration in Europe's demographic future has elicited concerns about the capacity of European societies to fully integrate and assimilate immigrants, and the long-term impact of immigrants on the ethnic and cultural identity of the host societies (Coleman 2006). The need for migrant labor in the face of increasing public anxiety and opposition to continued immigration has spurred renewed interest in temporary-migration programs, which are increasingly viewed by policy makers as beneficial to both host and source countries (Amin and Mattoo 2005; Walmsley and Winters 2005; Winters et al. 2003). Such programs provide host countries the labor they need without the social costs of immigrant incorporation, and they provide source countries the saved earnings and enhanced human capital of returning migrants. The success of temporary-migration programs in meeting the expectations of both host and source countries hinges not only on migrants returning home at the end of their contracts, but also on the amount of savings and new skills they bring back (IOM 2010; Plaza 2008). This chapter uses retrospective occupational and migration histories collected in Mexico to examine the occupational experiences of Mexico-U.S. migrants after returning to Mexico. The North American case shares many parallels with contemporary migration patterns in Europe and can highlight factors that influence the transferability of financial

D. P. Lindstrom (✉)
Population Studies and Training Center, Brown University, Providence, RI, USA
e-mail: David_Lindstrom@brown.edu

G. Neyer et al. (eds.), *The Demography of Europe*,
DOI: 10.1007/978-90-481-8978-6_8,

and human capital acquired from migration to source country labor markets—a key element of the current rationale for temporary-migration programs.

The chapter examines the impact of the event of return on occupational transitions, and the impact of cumulative U.S. migration experience on occupational transitions and life-time occupational mobility in Mexico. Results from the analysis suggest that the risk of downward occupational mobility at the time of reentry into the Mexican labor market is an added cost to returning home that could discourage return migration. Our results also shed light on the use of remittances for the purchase of land and the establishment of businesses in home communities. In addition to being an avenue for upward economic mobility, investments in capital assets and self-employment may be a substitute for employment in the home country labor market.

## 8.2 Background: Guest-Worker and Temporary-Migration Programs

### *8.2.1 The European Experience*

Economic recovery and growth in Northwestern Europe after the end of the Second World War created a demand for manual labor that could not be satisfied by the domestic labor force alone. Beginning in the early 1960s, the Federal Republic of Germany and other European countries partially met the demand for workers through the establishment of guest-worker programs with Turkey and later other countries. The oil crisis in 1973 and the accompanying economic slowdown led to the termination of most guest-worker programs (Ünver 2006). The original intention of the guest-worker programs was that migrants would return to their home country after their contracts ended (Coleman and Rowthorn 2004). Many workers did return home, but many remained and were joined by family members. Sluggish economic growth in the 1980s and growing unemployment led to efforts by some European countries to encourage former guest-workers and their families to return to their home countries. The largest importers of guest-workers, Germany and France, initiated repatriation schemes that provided financial incentives to immigrants to return to their countries of origin (Dustmann and Kirchkamp 2002; Fassmann and Munz 1992). Relatively few immigrants took the incentives. The flow of migrants increased in spite of efforts by countries to tighten their borders (Hollifield 1994; Hooghe et al. 2008; Jandl 2007).

In addition to the gradual growth of immigrant stocks in many Northwestern European countries, population aging and the demand for low-skilled workers transformed former labor exporting countries such as Ireland, Portugal, Spain, Italy and Greece into countries of immigration (Arango and Martin 2005; Marques 2010; Peixoto 2009). The expansion of the European Union into Central and Eastern Europe in recent years has been an important stimulus for migration from former

East Bloc countries into Northwestern and Southern Europe (Ruhs and Anderson 2010). Poles, Albanians, and Ukrainians can now be found in large numbers in countries like the UK, Portugal, and Italy (Kosic and Triandafyllidou 2004). Immigrant populations from Africa, Asia, and Latin American are also sizeable in many European countries and constitute approximately 60 % of all non-EU foreign born residents in the EU-27 countries (Vasileva 2010). As of 2009 the percent of foreign born residents in many European countries was comparable to, or higher than, the percent of foreign born residents in the United States (12.2 %) (U.S. Census Bureau 2009: Table 1.1). For example, the percent foreign born in Ireland was 14.1, in Germany 11.6, in France 11.0, in Spain 11.1 and in Denmark 8.8 (Vasileva 2010: 2). With the rise in human smuggling, irregular migration is increasingly viewed as a serious problem in the European Union. A number of responses at the national and EU level have been implemented to try to discourage irregular migration including employer sanctions, stronger enforcement efforts, and exclusionary social policies (Engbersen and Broeders 2009). There is renewed interest in Europe in returning to temporary-migration programs as a way to meet the demand for labor without resorting to permanent immigration (Commission of the European Communities 2005; Hoekman and Özden 2010; Martin 2006; Ruhs and Martin 2008). Several EU countries have established Mobility Partnerships and other programs granting temporary work visas with countries on the outer borders of the EU based on the idea that migration can be managed (Castles 2006; Djajić and Michael 2009; Finotelli and Sciortino 2009; Parkes 2009). The underlying assumption for many of the new temporary migration programs is that both the host and source countries can influence return migration through coordinated policies (Djajić and Michael 2009). Indeed, many migrants from new EU countries as well as countries on the periphery of the EU return to their home countries, and some engage in a pattern of repeated migration, contributing to the renewed confidence that migration flows can be effectively managed (Barrell et al. 2010; Constant and Zimmermann 2011; Rye and Andrzejewska 2010).

### *8.2.2 The North American Experience*

One of the largest cross-border migration systems in the world is Mexico-U.S. migration. As of 2009 an estimated 11.5 million Mexican born persons resided in the United States (Grieco and Trevelyan 2010: 2). About 58 % of this population is estimated to be in the United States without legal documentation (Passel and Cohn 2011: 11). Although migration from Mexico to the United States has a history reaching back to the nineteenth century, the evolution of current migration patterns can be traced to the Bracero program. The Bracero program was initiated by the United States with Mexico in 1942 to address labor shortages in agriculture created by military conscription during the Second World War. The program was repeatedly renewed over the years and by the time of its termination in 1964 a total of 4.6 million Mexican workers had been admitted into the United States on temporary work visas (Calavita 1992: 218). The actual number of Mexican workers entering the United States during the 22 year history of the Bracero program was much larger due to the undocumented migration that the program generated. Having established personal ties to employers in the United States, many migrants returned in subsequent years without authorization from the program. The termination of the Bracero program did not lead to the end of Mexico-U.S. migration, but rather ushered in a new era of undocumented migration (Massey et al. 2002).

By the mid 1980s, Mexico-U.S. migration had evolved into system of circular flows in which most migrants entered the United States without legal documentation; worked in agriculture, construction, and other seasonal jobs; and returned to Mexico within five years (Massey et al. 2002; Reyes 2001). The largest component of the circular flow was undocumented men, who tended to remain in the United States for durations of less than one year (Reyes 2001). The 1986 Immigration Reform and Control Act changed the dynamics of Mexico-U.S. migration. It provided a pathway to legalization for millions of undocumented immigrants in the United States, and at the same time it increased efforts to restrict unauthorized border crossings and the hiring of undocumented migrants. Both measures contributed to a trend toward longer migrant trips, and family reunification and settlement in the United States (Alba 2010; Massey et al. 2002).

During the most recent decade migration from Mexico to the United States has been marked by an overall decline in the flow of new entries, with the sharpest decline occurring after 2006 (Passel and Cohn 2009, 2010). Although the majority of Mexican migrants entering the United States in any given year continue to enter without legal documentation, the number of migrants entering on temporary work visas has been on the rise (Papademetriou et al. 2009). In 2009, 206,000 temporary work visas were issued for seasonal agricultural and non-agricultural work of which 179,000 went to Mexican workers (Department of Homeland Security 2009). The expansion of temporary worker programs is closely linked to current discussions in the United States of regularizing the status of many immigrants who do not have legal documentation, along with proposals for stepped-up interdiction efforts at the border and controls in the workplace on the hiring of unauthorized migrants.

A common lesson from the European and North American experiences with temporary worker programs is that they are not always temporary: in many instances they lead to the long-term settlement of migrant workers and their families (Jacoby 2003; Massey and Liang 1989). The current renewed interest in temporary-migration programs in Europe and the United States is in large part a response to the gradual, but steady, accumulation of settled immigrants from the earlier guest-worker programs. In spite of the renewed optimism that migrant labor flows can be managed with coordinated policies, very little is known about whether return migrants are able to successfully reincorporate themselves into home country labor markets, and the long-term returns on temporary migration experience in the home country. A better understanding of the occupational trajectories of return migrants can help to identify sources of friction to return migration flows, and policies that can reduce that friction.

## 8.3 Theory

### 8.3.1 Temporary Labor Migration and Occupational Mobility

Theories of temporary labor migration are either silent on the short and long-term consequences of temporary migration for occupational mobility in the place of origin, or they only consider the role of migrant savings in making the transition into business or land ownership. As we shall argue below, and as the results of our analysis suggest, there are reasons to believe that expectations regarding occupational outcomes after return may influence decisions about trip durations and eventual return. Most analyses and discussions of temporary labor migration draw upon one or more of three theoretical approaches: target income theory, household survival theory, and the New Economics of Labor Migration theory (NELM).

Target income theory was initially developed to explain temporary labor migration in Africa, and has been extended to the case of international labor migration to industrialized countries (Berg 1961; Hill 1987; Piore 1979). The theory assumes that workers have a strong preference for remaining in their home community, but because of limited local opportunities to earn cash they must resort to temporary labor migration. Optimally, migrants would prefer to spend as little time as possible away from home and yet accumulate enough savings from migrant earnings to reach a particular savings target (Berg 1961) or a desired level of net lifetime income (Hill 1987). The theory predicts that migrants return back to their place of origin after they reach a savings target that is based on specific consumption needs or investment plans in the home community. The theory implicitly views temporary labor migration as supplemental to locally produced goods and income, but is silent on how migration experience impacts earning capacity at home.

Household survival theory also views temporary labor migration as supplemental to local income, but it treats labor activities in the home community as part of an integrated household strategy to adapt to changing income needs and opportunities (Konseiga 2006; Wood 1981). It begins with the assumption that households are economic satisfiers that allocate the labor of household members in a way that meets basic consumption needs. In rural areas with limited sources of off-farm income, and in urban areas with restricted access to stable employment, households use the temporary labor migration of one or more of their members as a way to supplement locally generated income streams. While labor migration is used to cover household income deficits it does not alter long-term income generating capacity in the place of origin, and in some instances it may create a lasting dependence on migrant income to meet revised consumption needs (Reichert 1981). The theory assumes that migrants can smoothly transition back into income generating activities in their home community upon return from the place of destination and continue with those activities up to the timing of a new trip. Circular migration systems in which migrants alternate between work in their home community and temporary work in a place of destination have been documented in many diverse contexts and are consistent with the view of migration as supplemental to income earning activities in the place of origin (Hugo 1982; Massey et al. 1987; Oucho 1998).

The New Economics of Labor Migration also treats households as production and consumption units in which individual members pool resources and risk. The theory focuses on how the absence of smoothly functioning capital markets influences migration behavior (Stark 1991; Taylor 1999). One important market that is absent or poorly developed in low income countries, and especially in rural areas, is the credit market. The absence of affordable credit means that even in economically dynamic areas with investment opportunities, households may use migration to accumulate savings as a substitute for credit (Lindstrom 1996; Lindstrom and Lauster 1999). Research from a variety of different contexts document how households use migration to accumulate savings for investments in agricultural land and livestock, to capitalize small businesses, or for large lump sum purchases such as a house or a motor vehicle (León-Ledesma and Piracha 2004; Massey and Parrado 1998; Woodruff and Zenteno 2007).

Target income theory and household survival theory were developed to explain migration largely in the context of rural households, and the New Economics of Labor Migration demonstrates how temporary labor migration can be used as a vehicle for making capital investments at home. None of the three theoretical approaches to temporary migration address the re-incorporation of returning migrants into non-agricultural employment and in particular urban labor markets. Whether returning migrants are actually able to smoothly transition back into the occupations they held in their home communities prior to migrating is an open question. The failure to transition back to pre-migration income activities after return creates an element of risk and uncertainty in the process of return, and thereby imposes additional opportunity costs to temporary migration in the form of lost income in both the place of destination and the place of origin.

Results from prior research on the economic returns to temporary migration in place of origin are mixed. In a study of the income returns on foreign work experience in Hungary, Co et al. (2000) find that foreign work experience raises women's earnings but not men's. They suggest that the sectors in which men work after return to Hungary, mainly manufacturing and construction, do not reward foreign experience whereas the sectors in which women work, mainly financial services, reward foreign experience. They also suggest that the loss of contacts that men experience while working abroad may result in lower wages. Carletto and Kilic (2009) find in Albania that work experience in Italy and other more distant countries improves the chances of upward occupational mobility, whereas work experience in Greece has no impact on mobility. They attribute this differential effect of migration experience to the type of work that Albanian migrants typically perform in different destinations. Most Albanian migrants to Greece are engaged in agricultural and low-skilled work, whereas migrants to Italy and other countries are engaged in a greater variety of activities. In a study of Egyptian return migrants, McCormick and Wahba (2004) find that urban-origin migrants were more likely to report having acquired useful skills while working outside the country than rural-origin migrants. They also find that foreign work experience is associated with the accumulation of financial capital and investment in small businesses. Similarly, Ilahi (1999) in a study of Pakistani return migrants and Coulon and Priacha (2005) in a study of Albanian return migrants find that migrants with large savings from foreign work tend to opt for self-employment and establish small businesses. Two common themes in studies of the returns on foreign work experience are: (1) the type of employment in the destination country influences the opportunities to acquire valuable skills; and (2) foreign work experience is associated with self-employment and small business formation upon return to the home country. Some studies also report higher levels of unemployment among return migrants compared to pre-migration levels (Azam 1991).

### 8.3.2 Reentry into the Labor Market as a Type of Job Search

For theoretical and analytical purposes we identify two components of the relationship between international migration experience and occupational mobility in the country of origin. The first component is the impact that withdrawal from the origin labor market has on the chances that one is able to reenter the same occupation upon returning to the community of origin. The mere fact that one has left a job to out-migrate places one at risk of not returning to the same occupation that one held prior to migration. This potentially disruptive component of migration is present regardless of how international migration experience is valued in the origin labor market. The second component of the relationship between international migration experience and occupational mobility is the economic return in the home country on cumulative migration experience. International migrants may acquire skills while working abroad that enhance productivity and that are valued in the origin labor market. Migration experience can impact occupational transitions in the home community at the time of return, and can have a cumulative effect across migrants' working careers that impacts life-time occupational mobility.

### 8.3.3 Return Migration and the Job Search Process

To explain the process of reentry into the origin labor market we present a simple job search model. Let us assume that currently employed workers compare wage offers to their current wage and transition to a new job when a wage offer is better than their current wage. In this case we would expect voluntary job transitions to result in a wage that is equal to, or better than the current wage. In the absence of a better wage offer, employed workers remain in their current job (Parsons 1973). In the case of unemployed workers, wage offers are compared to a reservation wage, which is the lowest wage a worker is willing to accept. Unemployed workers remain unemployed until they locate a wage offer that is above the reservation wage (Lippman and McCall 1976). There is an opportunity cost, in the form of lost wages, associated with remaining unemployed during the job search (Mattila 1974; Mortensen 1986). There are also diminishing returns in the job search process. As the most promising and desirable job potentials are exhausted, the search is extended to increasingly less desirable jobs and more distant labor markets, which places downward pressure on the reservation wage (Kasper 1967).

In both the case of employed and unemployed workers, the expected waiting time to a better wage offer and a job transition is determined by the worker's skills, experience, and age; labor market conditions including turn-over and growth in positions for which the worker's human capital endowments are a good match; and the extensiveness and labor market location of the worker's social networks (Montgomery 1991). There is also a stochastic component in the occurrence and timing of better wage offers. From the perspective of an employed or unemployed worker the waiting time until a desirable job becomes available is not predictable (Mortensen 1986; Van Dijk and Folmer 1985).

Let us now extend this simple job search model to the case of return migration and occupational change. If the decision to return to the place of origin is linked to job offers in the place of origin, then the process of return migration and reentry into the origin labor market is similar to the situation of the currently employed. Temporary migrants in a destination who have met a savings target, begin the search for employment in their place of origin while they are still working in the place of destination and time their return to a job offer that is comparable or better to the last job they held in the place of origin before migrating to the destination country. In this case we expect:

1. Return to be associated with reentry into a similar or better paying occupation.

The timing of return migration in connection to securing a job that is comparable or better to the job that one left prior to migration minimizes the lost income that is associated with the job search process after return. It also maximizes the total amount of migrant savings that can be used for capital investments or non-essential expenditures by establishing an income stream in the place of origin as soon as possible.

If the decision to return is not linked to job offers in the place of origin, then the process of return migration and reentry into the origin labor market is similar to the job search process of the unemployed. The event of returning to the origin labor market is equivalent to being involuntarily separated from a job in the sense that the state of being without employment and engaged in the job search is not timed to existing wage offers. Because of both the stochastic component in the timing of desirable job offers and the diminishing returns in the job search process we expect that:

2. Return migration is associated with both transitions into a better paying and a worse paying occupation than the one held prior to out-migration from the place of origin.

Because being unemployed while engaged in the search process is associated with the depletion of migrant savings there is a strong incentive for return migrants to accept a less desirable job (or lower occupation) in the absence of a comparable or better job offer. Therefore, we expect:

3. The risk of returning to a worse paying occupation to be larger than the risk of returning to a better paying occupation.

This last hypothesis is consistent with the observation made by Blau and Robins (1990) that employed workers are more successful in the job search process than unemployed workers, and therefore, unemployed workers might improve their chances in the search process if they accept the first offer available and then continue the job search as an employed worker. The pressure on returning migrants to take the first job available is also consistent with negative duration dependence in the time it takes the unemployed to find a job—the longer the unemployed search for a job the less likely they will find one (Pissarides 1992).

### *8.3.4 The Returns on Cumulative Migration Experience in the Place of Origin*

The second component in the relationship between international migration experience and occupational mobility is the return on cumulative migration experience. We identify three possible mechanisms that link cumulative migration experience with occupational mobility in the place of origin. The first mechanism is migration as an investment in human capital. Employment in more economically advanced labor markets may provide migrants with new skills and work habits that enhance their productivity in the origin labor market. Many migrants also acquire experience in small service and manufacturing establishments such as restaurants, automobile and machinery repair shops, and metal fabricating that when combined with migrant savings can be used to establish a business in the place of origin. Under the human capital approach the impact of work experience in the destination labor market on occupational mobility in the place of origin will depend on the transferability of skills acquired in the destination labor market to the origin labor market, the extent to which migrant work experience actually enhances productivity, and the extent to which employers in the origin labor market value migration experience. If migration is an investment in human capital that enhances productivity and occupational mobility in the place of origin, then we expect:

4. Cumulative migration experience to improve the chances of upward skill-based mobility through the acquisition of new skills.
5. Nonagricultural work in the country of destination to have a bigger impact on upward mobility than farm work.

The second mechanism linking cumulative migration experience to occupational transitions in the place of origin is the use of migrant savings for capital investments. This mechanism is described by the New Economics of Labor Migration and it directly links the duration of cumulative experience to the total amount earnings that migrants are able to save and remit back home for investment purposes. If migration is a substitute for credit, then we expect:

6. Cumulative migration experience to improve the chances of movement into agricultural land or business ownership in the place of origin.

Because nonagricultural wages tend to be higher than agricultural wages, we expect:

7. Cumulative nonagricultural work to have a bigger impact on movement into agricultural land or business ownership than farm work.

The third mechanism linking migration experience to occupational transitions is the disruptive impact of being away from the origin labor market. While they are away from home, temporary migrants experience some deterioration in their origin-specific human capital. As migrants spend more time in a destination labor market and interact with fellow migrants, work mates, and employers, they build-up the quality and breadth of their social networks and job connections in the destination labor market, yet they also neglect their social and employment networks in the place of origin. The drop in investments in origin network connections reduces the quality of information about job-openings that returning migrants can expect to receive, and the quality of references and recommendations that migrants can expect to receive from former employers. If the skills acquired while working in the destination labor market are not transferable to the place or origin, or if few or no new skills are acquired, then migrants will also lose out on the place-specific skills and experience they would have acquired in the place of origin if they had not migrated. Returning migrants, therefore, experience some depreciation in the value of the work experience they accumulated in the place of origin before migrating and are not able to compensate for this depreciation with the experience they accumulated in the place of destination. Employers in the place of origin may also discriminate against return migrants because they view temporary migrants as being weakly attached to the origin labor market or they may view return as a sign of failure in the destination labor market (Barrett and O'Connell 2001; Schwab 1999). In either case, employers may be reluctant to hire return migrants or to make on-the-job investments in their training. The expected deterioration in origin-specific human capital associated with temporary migration and the potential employer penalty on migration experience leads to the expectation that:

8. Cumulative migration experience has no effect or a negative effect on upward skill-based mobility, and increases the chances of downward mobility, with no difference in the effects of nonagricultural and agricultural experience.

## 8.4 Data and Methods

For the analysis, we use retrospective life-history data collected by the Mexican Migration Project for male household heads in 88 Mexican communities. The communities are drawn from 17 of the 32 Mexican states, and incorporate traditional migrant sending regions and relatively new source areas of migration to the United States. The communities were purposively selected to represent a range of sizes, economic bases, and migration levels. They encompass villages and secondary towns, market towns, cities, and metropolitan areas. In most communities the sample consists of 200 households selected through simple random sampling, although samples tended to be smaller in the less populated places. Sampling frames were constructed by conducting a census of all dwellings in the community, or of specific neighborhoods in the case of large urban areas. Interviews in Mexico were typically conducted in December and January, when the return of migrants to Mexico for the Christmas holidays is at a peak. Interviews were conducted with the household head and spouse of the head if the household was headed by a couple. In cases where the household head was away in the United States at the time of the survey, the spouse or another senior member of the household provided data on the household head's migration and occupational history. Data for the 88 communities were collected between 1987 and 2002, with 3–6 communities surveyed in most years (http://mmp.opr.princeton.edu).

The occupational histories record change in occupations (not jobs), and change in place of occupation at the municipal, state, and country level. We use the retrospective migration and occupational histories for male household heads ages 25 and above to construct a life history file in which each record represents one life year. The life-histories start at age 17 or the age at first occupation if the household head started working after age 17, and are right-censored at the year of the survey, or at age 65, or in the last year of economic activity for men who became disabled or retired before age 65. Complete migration and occupational histories were available for 9,356 male household heads.

We classified the occupations into five categories: farm workers with 10 hectares or less of farm land, unskilled, skilled, professional, and land (more than 10 hectares) or businesses owners. Street vending and market stalls were not considered businesses. We treat the farm workers, unskilled, skilled, and professional occupations as ordered categories from lowest to highest based on education and skill requirements, and earnings. An analysis of the mean incomes for household heads in these categories validated this ordering. Although the mean income for land and business owners was greater than that of professionals, we treat the land and business owner category as an unordered occupational category. We defined occupational transitions as a change in occupational category from one life year to the next. Because we are interested in occupational change in Mexico, and in particular occupational change among return U.S. migrants, we assign the occupation a migrant held in Mexico prior to departure on a U.S. trip to the life years during which the migrant was in the United States. An occupational transition upon return to Mexico occurs if the occupation the migrant enters after return is different from the occupation held prior to migrating to the United States. Because the time unit used for the occupational histories is a year, unemployment spells of less than one year are not recorded. The year in which an unemployment spell occurs is classified by the primary occupation during the year or the last occupation held prior to unemployment.

Figure 8.1 presents a graphical image of a migration and occupational history. At age 17 and year $t_1$ the subject is in an unskilled occupation. In year $t_a$ the subject migrates to the United States and in year $t_b$ the subject returns to Mexico. Upon return to Mexico the subject enters into a skilled occupation and thus experiences an upward skill-based occupational transition. In year $t_c$ the subject experiences a downward occupational transition into an unskilled occupation, and remains there until the year of the survey. The life years $t_1$ to $t_b$ constitute an occupational spell during which there is no change in the occupational category in Mexico. The life years $t_b$ to $t_c$ constitute a second occupational spell during which the subject remains in a skilled occupation, and the years $t_c$ to $t_{\text{survey}}$ constitute a third occupational spell that ends in censoring. In this example the subject contributes $t_{\text{survey}}-(t_1-1)$ life years and three occupational spells to the occupational transition analysis file.

We use hazard regression models to estimate the impact of return migration and cumulative migration experience on the likelihood of making upward and downward skill-based occupational transitions, and transitions into land or business

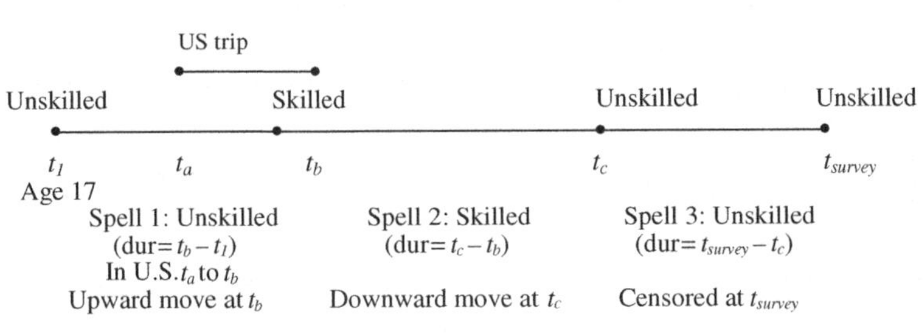


**Fig. 8.1** Occupational spells and life-time mobility

ownership. The conditional hazard of an occupational transition in year $t_i$ is defined as:

$$h_i(\mathrm{t_i}|\mathrm{X}, \theta) = \exp[\beta_{i0} + \mathrm{X}(t_i)\beta_i + \gamma_i(t_i^{\gamma} - 1) + c_i\theta]$$

where $\beta_{io}$ is a constant term, $\mathrm{X}(t_i)$ is a vector containing the values of the independent variables at time $t_i$, $\beta_i$ is a vector of coefficients, $\gamma_i$ $(t_i^{\gamma}-1)$ is a term for modeling Weibull duration dependence, and $c_i\theta$ is a non-parametric correction term for unobserved heterogeneity specified by Heckman and others (Flinn and Heckman 1982; Heckman and Singer 1984).

We estimate three transition models: the first model estimates the risk of an upward skill-based transition, the second model estimates the risk of a downward skill-based transition, and the third model estimates the risk of a transition into land or business ownership. In each model the risk of a transition is compared to making no transition, and the occupational spell is right-censored if the spell ends in a transition different from the type of transition that is being examined (e.g. the occurrence of a downward transition in the model estimating the risk of an upward transition), or the spell ends at age 65 or the year of the survey.[1] We do not model the risk of transitioning out of land or business ownership because very few individuals move out of this category.

After analyzing occupational spells, we look at the impact of cumulative migration experience on life-time occupational mobility in Mexico using the first and last occupational spells in the occupational history file. For the analysis of life-time occupational change we define three occupational starting points (farm worker, unskilled/skilled, and professional) based on first occupation. We then use multinomial logistic regression models to estimate the impact of cumulative migration experience on the likelihood of being in a different occupational category (downward mobility, upward mobility, or land/business ownership) at age 65 or in the year of the survey (the last observation point). We estimate separate models for each starting point because the set of possible outcomes varies. Men starting as farm workers can only experience upward mobility into an unskilled, skilled, or professional occupation, or mobility into land or business ownership. Men starting in unskilled or skilled occupations can experience upward mobility, downward mobility, or mobility into land or business ownership; and professionals can experience downward mobility or mobility into land or business ownership. Similar to the case of occupational spells, we do not model life-time occupational mobility out of land or business ownership because relatively few men start their working lives in this category, and very few men experience mobility out of it.

[1] Professional occupation spells are excluded from the model of upward mobility because there is no occupational category above professional. Similarly, farm worker spells are excluded from the model of downward mobility because there is no occupational category below farm worker.

## 8.5 Descriptive Statistics

Table 8.1 presents selected descriptive statistics for the sample of male household heads. The characteristics presented in the table correspond to the year of the survey or age 65. The study subjects are drawn from birth cohorts spanning almost 44 years and their working years cover a half century of change in the Mexican economy and U.S. immigration policy.

On average men in the sample have 6.5 years of completed schooling, although the spread in the distribution is substantial. Roughly one-in-three men had three or fewer years of completed education and another one-in-three men had some secondary education or beyond (figures not shown). Entry into marriage or consensual unions for men and women in Mexico is nearly universal. By age 65, or the

**Table 8.1** Selected descriptive statistics, male household heads. Mexican migration project, 1987–2002

| | Percent | Mean (std dev.) |
|---|---|---|
| *Background characteristics* | | |
| Cohort 1933–1939[1] | 12.7 | |
| 1940–1949 | 25.7 | |
| 1950–1959 | 29.9 | |
| 1960–1977 | 31.7 | |
| Period (life years)[2] | | |
| 1950–1964 | 8.8 | |
| 1965–1981 | 37.6 | |
| 1982–1989 | 28.1 | |
| 1990–2002 | 25.5 | |
| Years of schooling[1] | | 6.5 (4.7) |
| Married[1] | 96.4 | |
| *Migration experience*[1] | | |
| U.S. experience | 35.2 | |
| U.S. farm experience | 16.8 | |
| U.S. nonfarm experience | 23.6 | |
| U.S. legal documents | 7.3 | |
| Internal migrant | 41.6 | |
| *Family of origin U.S. migration experience*[1] | | |
| Parents U.S. migrants | 14.4 | |
| Siblings U.S. migrants | 39.9 | |
| *Community context*[1] | | |
| Prevalence of male U.S. migration | | 0.29 (0.17) |
| Rural village | 19.6 | |
| Town | 24.4 | |
| City | 27.9 | |
| Metropolitan area | 28.1 | |
| Number of observations = 9,356 | | |

*Note* [1] Corresponds to year of survey or age 65, [2] Percentages based on all life years in occupational transition file

year of the survey, slightly more than 96 % of the men were in a marital or consensual union.

U.S. migration experience is relatively common in the sample with around one-third of the men having worked in the United States. Close to 17 % of men worked in the United States in agriculture, and around 24 % worked in non-agricultural jobs. Possession of immigration documents permitting employment and entry at will was less common in the sample—only 7 % of men possessed U.S. citizenship or a residency card. Internal migration experience is also fairly common in the sample. Around 40 % of the men either migrated temporarily in Mexico to work in another location, or moved away from their place of birth to another community. An important facilitator of U.S. migration behavior is being socially connected to other experienced migrants. Close to 15 % of the men had at least one parent with U.S. migration experience and 40 % had at least one sibling with U.S. migration experience. At the community level, men in the sample on average lived in communities where roughly 30 % of the adult men had been to the United States at least once.

Figure 8.2 presents a life-time occupational mobility table with first occupation presented in the rows and the last or most recent occupation presented in the columns. The shaded cells along the main diagonal correspond to men who were in the same occupation at the time of last observation that they held at age 17. The cells above the main diagonal correspond to upward occupational mobility and the cells below the main diagonal correspond to downward mobility. The column and row corresponding to land or business ownership are set apart from the rest of the table to indicate that we treat this occupational category as unordered.

| | *Last Occupation* | | | | | |
|---|---|---|---|---|---|---|
| ***First Occupation*** | Farm work | Unskilled | Skilled | Professional | Land/busi. owner | Total |
| Farm worker | 19.4% | 4.8% | 5.6% | 0.9% | 6.9% | 37.6% |
| Unskilled | 1.7% | 10.7% | 8.6% | 1.6% | 7.1% | 29.6% |
| Skilled | 0.7% | 2.7% | 12.4% | 1.4% | 5.4% | 22.6% |
| Professional | 0.1% | 0.4% | 0.5% | 4.8% | 1.8% | 7.5% |
| Land/business owner | 0.1% | 0.1% | 0.1% | 0.0% | 2.3% | 2.5% |
| Total (Number of cases) | 22.0% | 18.6% | 27.2% | 8.8% | 23.5% | 100.0% (9,356) |

**Fig. 8.2** Life-time occupational mobility, male household heads, Mexican migration project, 1987–2002

Close to one-half of the men had yet to experience any life-time occupational mobility at the time of last observation, whereas close to one-in-four men (22.9 %) experienced upward skill based mobility and a slightly smaller percentage of men (22.1 %) had moved into land or business ownership. Only 6 % of men experienced downward life-time occupational mobility, and very few men who started out as land or business owners moved into a different occupation. The overall upward shift in the occupational distribution and the movement into land or business ownership reflects both the gradual industrialization of the Mexican economy that occurred during the lifetimes of the men in the sample, and the processes of skill-building and capital accumulation that occur within cohorts as part of the aging process. The two biggest occupational shifts across the life-course are the movements out of farm work and unskilled labor, and the movement into land or business ownership. At the outset of their working lives only 2.5 % of the men owned more than 10 hectares of agricultural land or a business. At the time of last observation 23.5 % of the men were land or business owners.

While only 6 % of the men in the sample had a last occupation that was less skilled than their first occupation, a much larger percentage of men experienced downward occupational mobility at some point in their work careers. Figure 8.3 presents an occupational transition table. The unit of analysis in this figure is an occupational spell. An occupational spell is defined as a continuous time period during which a subject remains in the same occupation. Forty-two percent of the men in the sample contribute only one occupational spell to the analysis, whereas 25 % contribute three or more spells (not shown in figure). The rows correspond to occupation at the start of a spell and the columns correspond to occupation at the end of the spell or the start of the next spell. The cells on the main diagonal correspond to spells that do not end with a transition into another occupation (right censored), and the off-diagonal cells correspond to occupational spells that end with a transition into a different occupation. Roughly one-in-eight (13.4 %) occupational spells end in a transition into a less-skilled occupation. Most of these transitions occur among unskilled and skilled workers who transition into farm work or unskilled occupations. Similar to what we saw in the case of life-time mobility, very few men who enter into land or business ownership transition out of this occupational category, suggesting that men who are able to accumulate capital assets are generally able to hold onto them.

In the next section we use multivariate regression models to estimate the strength and nature of the relationship between return migration and cumulative migration experience, and occupational change. We expect return migration to be associated with a higher risk of occupational change of any type. Temporary labor migration to the United States entails withdrawal from the Mexican labor market, which automatically places return migrants at risk of entering into an occupation different from the occupation they held prior to migration.

## 8.6 Occupational Transitions

We first analyze the occupational spells summarized in Fig. 8.3. We estimate separate Weibull hazard regression models for each of the three possible types of occupational transitions: upward mobility, downward mobility, and movement into land or business ownership. Spells starting in a professional occupation are excluded from the analysis of upward mobility because upward mobility out of the highest occupational group is not possible. Similarly, spells starting in farm work are excluded from the analysis of downward mobility. The spells are right censored if a transition other than the one being modeled occurs, or if the spell ends at age 65 or the year of the survey.

Table 8.2 presents parameter estimates from the hazard models predicting occupational transitions. The models include as background characteristics age and age-squared at the start of the occupational spell, years of completed education, occupation at the start of the spell, and marital status at the start of the spell. Measures of migration experience include a time-varying dummy variable indicating return from the United States in a given year, cumulative U.S. farm and nonfarm work experience at the end of the most recent U.S. migrant trip, a time-varying dummy variable indicating possession of U.S. legal residency or citizenship,[2] a dummy variable indicating U.S. migration status in the prior year, and

| *Occupation at start of spell* | Farm work | Unskilled | Skilled | Professional | Land/busi. owner | Total |
|---|---|---|---|---|---|---|
| | *Occupation at end of spell* | | | | | |
| Farm worker | 10.8% | 6.2% | 4.7% | 0.6% | 3.1% | 25.4% |
| Unskilled | 3.8% | 9.0% | 9.1% | 1.5% | 3.9% | 27.3% |
| Skilled | 2.3% | 5.3% | 13.2% | 1.8% | 4.1% | 26.6% |
| Professional | 0.2% | 0.7% | 1.1% | 4.3% | 1.6% | 7.8% |
| Land/business owner | 0.5% | 0.7% | 0.7% | 0.3% | 10.8% | 12.9% |
| Total (Number of spells) | 17.5% | 21.7% | 28.9% | 8.4% | 23.5% | 100.0% (18,569) |

**Fig. 8.3** Occupational transitions, male household heads, Mexican migration project, 1987–2002

[2] Legal documents includes legal resident (Green card), citizenship, Silva letter (special status given primarily to Mexican immigrants in the late 1970s that led to legal residency in the 1980s, and refugee or asylum status.

**Table 8.2** Parameter estimates from Weibull hazard regression models predicting occupational transitions, male household heads, life years ages 17–65, Mexican migration project, 1987–2002

| | Model 1 | | Model 2 | | Model 3 | |
|---|---|---|---|---|---|---|
| | Upward mobility | | Downward mobility | | Land/bus. ownership | |
| | $\beta$ | | $\beta$ | | $\beta$ | |
| *Background characteristics* | | | | | | |
| Age | −0.045 | *** | −0.062 | *** | 0.022 | |
| Age-squared | 0.001 | | 0.002 | *** | 0.000 | |
| Education | 0.114 | *** | −0.050 | *** | 0.092 | *** |
| Farm worker (ref. model 1, 3) | 0.000 | | N.A. | | 0.000 | |
| Unskilled occupation (ref. model 2) | −0.353 | *** | 0.000 | | 0.591 | *** |
| Skilled occupation | −2.892 | *** | 0.847 | *** | 0.327 | *** |
| Professional occupation | N.A. | | 1.084 | *** | 0.134 | |
| Married | 0.159 | *** | −0.055 | | 0.518 | *** |
| *Migration experience* | | | | | | |
| Return from U.S. | 2.441 | *** | 2.854 | *** | 1.009 | *** |
| U.S. farm exp. (log months) | −0.148 | *** | 0.011 | | 0.063 | ** |
| U.S. nonfarm exp. (log months) | −0.111 | *** | −0.068 | ** | 0.131 | *** |
| U.S. documents | −0.858 | *** | −0.564 | * | 0.039 | |
| $\text{Lag}_1$ in U.S. | −0.720 | *** | −0.909 | *** | −0.411 | *** |
| Internal migrant | 1.973 | *** | 1.417 | *** | 0.440 | *** |
| *Family of origin U.S. mig. exp.* | | | | | | |
| Parents U.S. migrants | 0.267 | *** | −0.000 | | 0.252 | *** |
| Siblings U.S. migrants | −0.041 | * | 0.016 | | 0.068 | *** |
| *Community context* | | | | | | |
| Prevalence of male U.S. migration | −0.546 | *** | 0.550 | ** | 0.387 | ** |
| Employment opportunity index | 0.185 | *** | −0.332 | *** | −0.114 | ** |
| Rural village (ref.) | 0.000 | | 0.000 | | 0.000 | |
| Town | 0.443 | *** | −0.522 | *** | 0.185 | ** |
| City | 0.537 | *** | −0.571 | *** | 0.119 | |
| Metropolitan area | 0.567 | *** | −0.377 | ** | −0.015 | |
| *Period and spell controls* | | | | | | |
| Pre-1965 (ref.) | 0.000 | | 0.000 | | 0.000 | |
| 1965–1981 | 0.141 | ** | 0.096 | | −0.286 | *** |
| 1982–1989 | 0.159 | ** | 0.231 | ** | −0.311 | *** |
| 1990–2002 | 0.103 | | 0.255 | * | −0.121 | |
| Spell 1 (ref.) | 0.000 | | 0.000 | | 0.000 | |
| Spell 2 | −0.372 | *** | 0.528 | *** | 0.012 | |
| Spell 3 | −0.421 | *** | 0.187 | | 0.099 | |
| Spell 4 | 0.156 | | 1.131 | *** | 0.278 | ** |
| *Model parameters* | | | | | | |
| Constant | −3.907 | *** | −3.179 | *** | −5.654 | *** |
| Duration dependence $\gamma$ | −0.019 | *** | −0.022 | *** | 0.037 | *** |
| Heterogeneity parameter $c$ | −2.569 | *** | −2.880 | *** | −2.727 | *** |
| Latent group proportion $p$ | 0.573 | *** | 0.338 | *** | 0.442 | *** |
| Likelihood ratio chi square | 3,224 | *** | 2,635 | *** | 1,067 | *** |
| Number of spells | 14,726 | | 11,456 | | 16,171 | |

*Note* ***$p < 0.01$, **$p < 0.05$, *$p < 0.10$; N.A., occupational group not included in analysis
N.A. = Spells in occupational category excluded from the model

a dummy variable indicating internal migration experience in any year prior to the current year. To measure migration social capital the models include time-varying dummy variables indicating whether either parent or any siblings were U.S. migrants in any prior year, and the prevalence of male U.S. migration in the home community at the start of the occupational spell. In addition to the prevalence of U.S. migration, the models include an index of employment opportunities in the home community, and the level of urbanization. The employment index is constructed from eight municipal-level measures of economic activity taken from the 1950–2000 Mexican censuses using principal components analysis.[3] The models also include control variables for period and spell number, and a non-parametric control for spell-specific unobserved heterogeneity.

The first column in Table 8.2 presents coefficients from the model predicting the hazard of upward mobility, the second column presents results for the hazard of downward mobility, and the third column presents results for the hazard of making a transition into land or business ownership. Turning first to the human capital measures, we find that as men age they are less likely to make a skill-based occupational transition of any type. Occupational transitions tend to happen at younger ages when men are less risk averse and more actively engaged in the search for the best occupational match to their skills and interests. Having a higher level of education increases the chances of upward mobility and decreases the risk of downward mobility. More educated men are also more likely to turn to land or business ownership as a primary occupation. Controlling for age, married men are more likely than single men to experience upward skill-based mobility and movement into land or business ownership.

The year of return from the United States is associated with a significantly higher risk of all three types of occupational moves. Returning migrants are 11 ($e^{2.441}$) times more likely to transition into a higher skilled occupation and 17 ($e^{2.854}$) times more likely to transition into a less skilled occupation compared to non-migrants and return migrants in subsequent years. As expected, the mere fact that return migrants are reentering the labor market places them at a very high risk of experiencing some type of occupational change. The fact that the risk of experiencing a downward change is even higher than the risk of an upward change suggests that the timing of return migration is not linked to having a job waiting in Mexico, and the condition of reentry is more comparable to involuntary separation

---

[3] The index of employment opportunities is constructed from the female labor force participation rate, the proportion of economically active females working in the service sector, the proportion of economically active females working in manufacturing, the proportion of economically active males working in the service sector, the proportion of economically active males working in the manufacturing sector, the proportion of economically active adults who are employers, the proportion of economically active adults earning more than twice the minimum wage, and the municipal population. Principal components analysis was used to construct a composite index of employment opportunities for each municipality in each of the six census years (1950, 1960, 1970, 1980, 1990, and 2000). Linear interpolation was used to estimate values of the index in the intercensal years, and the value of the index in 2000 was used for the years 2001 and 2002 in communities that were surveyed after 2000.

from a job than voluntary separation. The chance of moving into land or business ownership is comparatively smaller than making a skill-based transition, but still important. Returning migrants are 2.7 ($e^{1.009}$) times as likely to acquire farmland or start a business in the year they return than nonmigrant or return migrants in subsequent years. The substantially lower risk of a transition into land or business ownership in the year of return compared to the risk of a skill-based occupational transition is consistent with the high financial barriers to entry into ownership, and the absence of a stochastic component in the determinants of land and business ownership. Returning to become a land or business owner is not a matter of being lucky or unlucky, rather it requires substantial savings and is the result of long-term planning and strategic action.

Cumulative U.S. migration experience represents both time working in the U.S. labor market as well as time spent away from the Mexican labor market. Both farm work and nonfarm work experience in the United States are associated with a significantly lower chance of upward mobility in Mexico. This result is consistent with the existence of a stochastic component in the relationship between return migration and occupational change. Reentering the Mexican labor market increases the risk of all types of occupational moves, but the increased risk of upward mobility is not due to any positive valuation of U.S. experience in the Mexican labor market. In fact, there appears to be a penalty associated with spending long periods of time in the United States in the form of a lower chance of upward mobility.

U.S. farm experience has no significant effect on downward mobility, however, nonfarm work is associated with a significantly lower risk of downward mobility. Generally, nonfarm wages are higher than farm wages and therefore provide a greater opportunity for migrants to save earnings and send money home. Returning migrants with more experience in nonfarm work are likely to have more savings to draw upon during the job search process, which allows them to extend the search for a desirable job longer than returning migrants with less experience or experience in agriculture. As predicted, both farm and nonfarm experience in the United States is associated with a significantly greater likelihood of becoming a land or business owner in Mexico. Also as predicted, the effect of experience is larger for nonfarm work than farm work because of the greater earnings potential typically associated with nonfarm work.

Possession of legal residency or U.S. citizenship is associated with a significantly lower risk of making either an upward or a downward skill-based occupational move in Mexico. Having U.S. documents is generally associated with a strong residential and work attachment to the United States. Migrants, who have legal residency or citizenship in the United States and continue to maintain a residence in Mexico, typically use the Mexican residence for rest and relaxation and as an option for retirement. Their attachment to the Mexican labor market is weak or completely severed and therefore they have little or no risk of occupational change in Mexico.[4]

Prior internal migration experience in Mexico is associated with an increased risk of all three types of occupational moves. Men migrate internally to take advantage of better occupational and investment opportunities available in other locations. However, they also migrate in search of employment if they experience employment loss in their place of origin.

Having a parent or parents with prior U.S. migration experience is associated with a significantly higher likelihood of both upward skilled-based mobility and movement into land or business ownership. Given the well-established relationship between migration and the family life-cycle, most of the parents would have been migrants when the men in our analysis were children. Migrant parents may pass-on higher economic aspirations to their children than non-migrant parents that subsequently get translated into occupational mobility and capital accumulation. In contrast to parental experience, having siblings with U.S. migration is associated with a significantly lower risk of upward mobility. Men whose siblings are in the United States have weaker social network ties to the local labor market than men whose siblings remain in Mexico. These weaker ties mean they are less likely to have leads on job openings in Mexico than would be the case if all their siblings were in Mexico. Having siblings with U.S. migration experience is also associated with a higher risk of moving into land or business ownership. Men with fewer employment options in Mexico are more likely to establish a business as an alternative to wage employment.

The effects of the prevalence of U.S. migration in the community of origin on occupational change are consistent with the results for individual migration experience and the migration experience of siblings. Men who live in communities with a high prevalence of U.S. migration are less likely to experience upward occupational mobility and are more likely to experience downward mobility or movement into land or business ownership. In communities where migration is common, men are likely to have better leads on job opportunities in the United States than in Mexico. These weaker social ties to the origin labor market compared to the destination labor market translate into fewer opportunities for upward mobility and a greater risk of downward mobility after separation from a job. The weaker social ties to the local labor market that are associated with a high prevalence of migration are also associated with a greater likelihood of investments in farm activities or small businesses as an alternative to employment based mobility.

The results for the index of employment opportunities and the level of urbanization are consistent with expectations. Better employment opportunities are associated with a greater likelihood of upward occupational mobility and lower likelihoods of downward mobility or movement into land or business ownership. This result is consistent with the idea that land and business ownership functions as

[4] By definition, the occupation in the life history remains unchanged until an occupational transition is made in Mexico.

an alternative to attractive employment opportunities. The likelihood of experiencing upward occupational mobility also increases with the level of urbanization, and the likelihood of experiencing downward mobility tends to decrease with the level of urbanization.

Overall the results from the analysis of occupational transitions are consistent with U.S. migration being disruptive of occupational trajectories in Mexico. The chances of experiencing upward mobility upon return to Mexico are not linked to U.S. migration experience, but rather to the reentry of returning migrants into the Mexican labor market. The chances that return migrants experience downward occupational mobility at the time of return are even larger than the chances of experiencing upward mobility. Consistent with other studies on the use of remittance income, we find that returning U.S. migrants are more likely to purchase farm land or establish a business than nonmigrant or migrants in subsequent years. While migration certainly functions as a substitute for scarce credit as argued by the New Economics of Labor Migration, the results suggest that investment in land and businesses is also a substitute for poor employment opportunities, especially for returning migrants. In the next section we look at life-time occupational change to determine whether return migrants are eventually able to overcome the disruptive effects of migration on employment trajectories in Mexico or if the disruptive effects persist over time.

## 8.7 Life-Time Occupational Mobility

Table 8.3 presents the results from the multinomial logistic regression models predicting life-time occupational change. We estimated three models conditional on type of first occupation. Model 4 estimates the likelihood of upward mobility and movement into land or business ownership for men starting in agriculture. Model 5 estimates the likelihood of upward mobility, downward mobility, and movement into land or business ownership for men starting in unskilled or skilled occupations; and Model 6 estimates the likelihood of downward mobility or movement into land or business ownership for men starting in professional occupations. The results for education are consistent with what we found in the case of occupational transitions: more years of schooling are associated with a greater likelihood of upward mobility and a lower likelihood of downward mobility. Higher education is also associated with a greater likelihood of becoming a land or business owner among farm workers and unskilled and skilled workers, but not among professionals.

In the case of life-time occupational change, cumulative U.S. migration experience has no impact on upward mobility among unskilled and skilled workers, and a negative impact on upward mobility among farm workers. Even more telling, U.S. migration experience actually increases the likelihood of long-term downward mobility among men starting out in non-farm occupations. These results strongly suggest that men who work in the United States and return to Mexico are

**Table 8.3** Parameter estimates from multinomial logistic regression models predicting life-time occupational change, male household heads, Mexican migration project, 1987–2002

| | Model 4 | | Model 5 | | | Model 6 | |
|---|---|---|---|---|---|---|---|
| | First occupation agriculture | | First occupation unskilled/skilled | | | First occupation professional | |
| | Upward mobility | Land/bus. ownership | Upward mobility | Downward mobility | Land/bus. ownership | Downward mobility | Land/bus. ownership |
| | β | β | β | β | β | β | β |
| *Background characteristics* | | | | | | | |
| Age | −0.028 * | 0.056 ** | 0.045 *** | 0.028 * | 0.050 *** | 0.012 | 0.076 *** |
| Age-squared | −0.000 | −0.002 ** | −0.001 * | 0.001 | −0.001 ** | 0.002 | −0.000 |
| Cohort 1933–1939 (ref.) | | | | | | | |
| Cohort 1940–1949 | −0.083 | −0.250 | 0.119 | 0.488 | −0.375 * | 0.724 | 0.329 |
| Cohort 1950–1959 | −0.164 | −0.420 | 0.293 | 0.719 * | −0.503 * | 1.323 | 0.761 |
| Cohort 1960–1977 | −0.745 | −0.986 * | 0.466 | 1.122 ** | −0.802 ** | 1.520 | 1.066 |
| Education | 0.125 *** | 0.167 *** | 0.094 *** | −0.010 | 0.107 *** | −0.192 *** | −0.030 |
| Married | 0.120 | −0.096 | 0.382 ** | 0.132 | 0.513 ** | -0.753 | 0.745 |
| *Migration experience* | | | | | | | |
| U.S. farm exp. (log months) | −0.073 * | 0.077 ** | 0.006 | 0.063 | −0.022 | 0.188 | −0.132 |
| U.S. nonfarm exp. (log months) | 0.025 | 0.051 | 0.040 | 0.094 ** | 0.083 ** | 0.217 * | 0.175 * |
| U.S. documents | −0.547 ** | −0.275 | −0.787 *** | −0.384 | −0.743 *** | 0.131 | 0.575 |
| Internal migrant | 1.002 *** | 0.544 *** | 0.493 *** | 0.536 *** | 0.311 *** | 0.850 ** | 0.100 |
| *Family of origin U.S. mig. exp.* | | | | | | | |
| Parents U.S. migrants | −0.032 | 0.243 | 0.142 | −0.267 | −0.044 | 0.118 | 0.412 * |
| Siblings U.S. migrants | −0.004 | 0.086 ** | −0.011 | 0.009 | 0.059 ** | −0.008 | 0.041 |
| *Community context* | | | | | | | |
| Prevalence of male U.S. mig. | −0.895 | −0.665 | −0.618 | 0.637 | 0.004 | 0.497 | 0.165 |
| Employment opportunity index | 0.615 *** | 0.059 | 0.061 | −0.303 ** | −0.310 *** | −0.168 | −0.302 * |
| Rural village (ref.) | | | | | | | |

(continued)

**Table 8.3** (continued)

| | Model 4 | | Model 5 | | | Model 6 | |
|---|---|---|---|---|---|---|---|
| | First occupation agriculture | | First occupation unskilled/skilled | | | First occupation professional | |
| | Upward mobility | Land/bus. ownership | Upward mobility | Downward mobility | Land/bus. ownership | Downward mobility | Land/bus. ownership |
| | $\beta$ | $\beta$ | $\beta$ | $\beta$ | $\beta$ | $\beta$ | $\beta$ |
| Town | 0.643 *** | 0.421 * | −0.018 | −0.648 ** | −0.073 | 0.820 | 0.073 |
| City | 0.658 *** | 0.235 | 0.158 | −0.804 *** | 0.042 | 1.381 ** | 0.031 |
| Metropolitan area | 1.733 *** | 1.494 *** | 0.023 | −0.747 * | −0.077 | 2.326 *** | 0.664 * |
| *Model parameters* | | | | | | | |
| Constant | −1.413 ** | −2.002 *** | −2.407 *** | −2.270 *** | −1.420 *** | −2.000 * | −2.846 ** |
| Likelihood ratio chi square | 1,053 *** | | 1,141 *** | | | 249.7 *** | |
| Number of cases | 3,520 | | 4,892 | | | 706 | |

*Note* ***$p < 0.01$, **$p < 0.05$, *$p < 0.10$

unlikely to experience any wage returns on their U.S. work experience and in fact may be penalized for their experience. The penalty likely derives from the deterioration in location-specific human and social capital that migrants experience as a result of being away from the Mexican labor market. It is also possible that employers discount U.S. migration experience because they perceive return migrants as being at a higher risk of quitting in order to undertake another trip to the United States. As expected, return migrants who have spent more time in the United States are also more likely than others to purchase farm land or establish a business. These results for cumulative migration experience reinforce our interpretation of the significant effects of return migration on upward and downward occupational transitions as capturing the impact of reentry into the labor market rather than the valuation of migration experience by potential employers.

Consistent with what we found in the case of occupational transitions, possession of U.S. immigration documents is associated with a significantly lower chance of upward mobility, and a lower chance of movement into land or business ownership among unskilled and skilled workers. Having siblings with U.S. migration experience increases the chances of becoming a land or business owner, although, having parents with U.S. migration experience does not have a significant effect on upward mobility as it did in the case of occupational transitions. The prevalence of male U.S. migration in the home community also has no significant effect on life-time mobility even though it was associated with a higher risk of downward occupational transitions and lower chances of upward transitions and transitions into land or business ownership. The weaker effects of family and community U.S. migration ties on life-time mobility compared to occupational transitions suggests that return migrants may be able to overcome some of the mobility setbacks associated with separation from the Mexican labor market.

## 8.8 Conclusions and Discussion

We find no evidence to suggest that Mexico-U.S. migrants returning to Mexico are able to convert their experience working in the United States into upward occupational mobility in Mexico. Both in terms of occupational transitions and life-time occupational change cumulative U.S. experience does not increase the chances of upward mobility and in some situations it is associated with a lower chance of upward mobility. The absence of any skill-based occupational gains to U.S. migration experience could be due to any of the following reasons: Mexican migrants are concentrated in low skilled jobs in the United States and have few opportunities to acquire new skills; the skills they acquire are not fully transferrable to the Mexican labor market; and Mexican employers do not value U.S. work experience and may even discount it. In contrast to the lack of evidence of positive returns on experience, we find evidence that cumulative nonfarm work experience in the United States is associated with a higher risk of downward life-time occupational mobility. The flipside of time spent working in the United States is time spent away from the Mexican labor market which leads to a deterioration of location-specific human and social capital that is critical for locating better paying jobs. To the extent that there is an element of randomness in finding a better job, workers who spend time in the United States are also exposed to the risk of finding a better job for less time than comparable nonmigrant. Mortensen (1986) suggests that the earnings of workers rise with experience not because of the returns on experience but because workers with longer experience have had more time to find a higher paying job. Even if there were no deterioration of human and social capital in the home country, just by being absent from the home country labor market temporary migrants have fewer opportunities over their life time to locate a better job than non-migrants.

Higher levels of cumulative U.S. experience are associated with an increased likelihood of purchasing farmland or establishing a business in Mexico. Prior research has linked migrant investments in capital assets to the expected returns on investments in the community of origin. The New Economics of Labor Migration theory predicts that migration is used to accumulate savings for investments in family enterprises when credit is scarce. Our result is consistent with this position, but we suggest that another reason for investment in land and businesses is the lack of attractive employment opportunities for returning migrants. The disadvantaged position of returning migrants in their origin labor market makes self-employment through business formation an attractive alternative to wage employment. We find that stronger parent and sibling ties to the United States and consequently weaker ties to the Mexican labor market are also associated with a higher likelihood of moving into land or business ownership. Consistent with our predictions, nonfarm work experience in the United States provides a better opportunity to accumulate savings for capital investments in Mexico than farm work because of the higher earnings associated with nonfarm work.

The biggest impact that U.S. migration has on the occupational trajectories of returning migrants is associated with reentry into the Mexican labor market. Some migrants are able to return to the work they held up to the time they migrated to the United States, this is especially the case with farmers and the self-employed. However, most migrants must locate new employment. We presented two alternative hypotheses regarding how return migrants reenter the home labor market based on a simple model of the job search process. The first hypothesis predicted that migrants began the job search process while still in the United States and timed their return to Mexico to finding a job that was comparable or better than the job they held in Mexico at the time of out-migration to the United States. The second hypothesis predicted that migrants did not time their return to having an attractive job offer in Mexico, in which case both upward and downward occupational transitions were likely to occur upon return to Mexico. Our results were consistent with the second hypothesis. We find that migrants are at a substantially higher risk of making any type of occupational transition in the year in which they return, and that the risk of making a downward skill-based transition was greater than the risk of making an upward transition. It is possible that there are migrants who time their return to Mexico to finding a desirable job, but the considerably higher risk of experiencing downward mobility at the time of return suggests that more returns are not timed to job offers in Mexico.

We suggest that the higher risks of both upward and downward mobility at the time of return reflects a component of the labor market reentry process that is random from the perspective of the returning migrant. The stochastic nature of finding a desirable job offer at the time of return introduces an important element of uncertainty and risk in return migration that increases the opportunity costs of return. In our review of the theories of temporary migration we noted that the different approaches were either silent on the return to economic activity or they assumed that return migrants smoothly transitioned into income earning activities. Our findings challenge this assumption. The risk of ending-up in a job that is worse than the job held prior to out-migration is likely to discourage return to Mexico among some migrants, and encourage self-employment among other migrants who decide to return to Mexico.

Our findings have implications for anticipating return migration among temporary low-skilled migrant workers in the United States and Europe. Temporary migrant programs function best when the work in the destination country is complementary to work in the origin labor market, as is the case of farm workers, or when the skills that are acquired in the destination country are transferrable and valued in the origin country. The early theoretical models of temporary labor migration were developed to explain the migration behavior of workers in rural, subsistence households where the transition between farm work and off-farm work was smooth. Similarly, the Bracero program between the United States and Mexico recruited workers from predominantly rural areas where reentry into the home labor market was non-problematic. In contrast to the earlier experiences of temporary labor migration, many of the source countries for current temporary labor migrants to the United States and to Europe are predominantly urban societies. Among temporary labor migrants from urban areas, withdrawal and separation from the home labor market and subsequent return and reentry come at a cost to long-term income streams in the home country. These costs have to be weighed against the saved earnings from temporary migration, and the expected earnings from staying on in the destination country beyond the termination of the work contract as an irregular migrant. Because much of the work that is targeted for temporary labor migrants in the United States and Europe is unskilled, the possibilities for acquiring skills that will be valued in the home country labor market are very limited. Consequently, in many instances the income rewards and incentives for return to the home country will be small. The design of sustainable temporary-migration programs needs to take into account the friction in the process of return migration that is generated by the costs of reentry into home country labor markets encountered by migrants engaged in low-skilled work.

## References

Alba, F. (2010). Mexico: a crucial crossroads. *Country Profiles February 2010*. Migration Information Source, Migration Policy Institute.

Amin, M., & Mattoo, A. (2005). Does temporary migration have to be permanent? Policy Research Working Paper Series WPS3582, World Bank.

Arango, J., & Martin, P. (2005). Best practices to manage migration: Morocco-Spain. *International Migration Review, 39*(1), 258–269.

Azam, F. (1991). Labour migration from Pakistan: Trends, impacts, and implications. *Regional Development Dialogue, 12*(3), 53–71.

Barrell, R., Fitzgerald, J., & Riley, R. (2010). EU enlargement and migration: Assessing the macroeconomic impacts. *Journal of Common Market Studies, 48*(2), 373–395.

Barrett, A., & O'Connell, P. J. (2001). Is there a wage premium for returning Irish migrants? *The Economic and Social Review, 32*(1), 1–21.

Bengtsson, T., & Scott, K. (2011). Population aging and the future of the welfare state: The example of Sweden. *Population and Development Review, 37* (Supplement), 158–170.

Berg, E. J. (1961). Backward-sloping labor supply functions in dual economies—The Africa case. *The Quarterly Journal of Economics, 75*(3), 468–492.

Bijak, J., Kupiszewska, D., & Kupiszewski, M. (2008). Replacement migration revisited: Simulations of the effects of selected population and labor market strategies for the aging Europe, 2002–2052. *Population Research and Policy Review, 27*, 321–342.

Blau, D. M., & Robins, P. K. (1990). Job search outcomes for the employed and unemployed. *Journal of Political Economy, 98*(3), 637–655.

Calavita, K. (1992). *Inside the state: The Bracero program, immigration, and the I.N.S.* New York: Routledge.

Carletto, C., & Kilic, T. (2009). Moving up the ladder? The impact of migration experience on occupational mobility in Albania. *Policy Research Working Paper* 4908, The World Bank, Development Research Group.

Castles, S. (2006). Guestworkers in Europe: A resurrection? *International Migration Review, 40*(4), 741–766.

Coleman, D. (2006). Immigration and ethnic change in low-fertility countries: A third demographic transition. *Population and Development Review, 32*(3), 401–446.

Coleman, D., & Rowthorn, R. (2004). The economic effects of immigration into the United Kingdom. *Population and Development Review, 30*(4), 579–624.

Coleman, D., & Rowthorn, R. (2011). Who's afraid of population decline? A critical examination of its consequences. *Population and Development Review, 37* (Supplement), 217–248.

Commission of the European Communities (2005). Migration and development: Some concrete orientations. COM(2005): 390.

Constant, A. F., & Zimmermann, K. F. (2011). Circular and repeat migration: Counts of exits and years away from the host country. *Population Research and Policy Review*, *30*(4), 495–515.

Co, C. Y., Gang, I. N., & Yun, M. (2000). Returns to returning. *Journal of Population Economics, 13*, 57–79.

de Coulon, A., & Piracha, M. (2005). Self-selection and the performance of return migrants: The source country perspective. *Journal of Population Economics, 18*, 779–807.

Department of Homeland Security (2009). Yearbook of immigration statistics; 2009. www.dhs.gov/files/statistics/publications/YrBh09N1.shtm.

Djajić, S., & Michael, M. S. (2009). Temporary migration policies and welfare of the host and source countries: A game-theoretic approach. CESifo *Working Paper No.* 2811.

Dustmann, C., & Kirchkamp, O. (2002). The optimal migration duration and activity choice after re-migration. *Journal of Development Economics, 67*, 351–372.

Engbersen, G., & Broeders, D. (2009). The state versus the alien: Immigration control and strategies of irregular immigrants. *West European Politics, 32*(5), 867–885.

Fassmann, H., & Munz, R. (1992). Patterns and trends of international migration in Western Europe. *Population and Development Review, 18*(3), 457–480.

Finotelli, C., & Sciortino, G. (2009). The importance of being southern: The making of policies of immigration control in Italy. *European Journal of Migration and Law, 11*, 119–138.

Flinn, C. J., & Heckman, J. J. (1982). New methods for analyzing individual event histories. In S. Leinhardt (Ed.), *Sociological methodology* (pp. 99–140). San Francisco: Jossey-Bass.

Grieco, E. M., & Trevelyan, E. N. (2010). Place of birth of the foreign-born population: 2009. *American Community Survey Briefs*, Oct 2010.

Heckman, J. J., & Singer, B. (1984). A method for minimizing the impact of distributional assumptions in econometric models for duration data. *Econometrica, 52*, 271–320.

Hill, J. K. (1987). Immigrant decisions concerning duration of stay and migratory frequency. *Journal of Development Economics, 25*, 221–234.

Hoekman, B., & Özden, C. (2010). The euro-mediterranean partnership: Trade in services as an alternative to migration? *Journal of Common Market Studies, 48*(4), 835–857.

Hollifield, J. F. (1994). Immigration and republicanism in France: The hidden consensus. In W. Cornelius, P. L. Martin, & J. F. Hollifield (Eds.), *Controlling immigration: A global perspective* (pp. 143–175). Stanford: Stanford University Press.

Hooghe, M., Trappers, A., Meuleman, B., & Reeskens, T. (2008). Migration to European countries: A structural explanation of patterns, 1980–2004. *International Migration Review, 42*(2), 476–504.

Hugo, G. (1982). Circular migration in Indonesia. *Population and Development Review, 8*(1), 59–83.

Ilahi, N. (1999). Return migration and occupational change. *Review of Development Economics, 3*(2), 170–186.

IOM (International Organization for Migration). (2010). *Managing return migration*. Geneva: International Organization for Migration.

Jandl, M. (2007). Irregular migration, human smuggling, and the eastern enlargement of the European Union. *International Migration Review, 41*(2), 291–315.

Jacoby, N. (2003). America's de facto guest workers: Lessons from Germany's Gastarbeiter for U.S. immigration reform. *Fordham International Law Journal, 27*(4), 1569–1662.

Kasper, H. (1967). The asking price of labor and the duration of unemployment. *Review of Economics and Statistics, 49*, 165–172.

Konseiga, A. (2006). Household migration decisions as survival strategy: The case of Burkina Faso. *Journal of African Economies*, 1–36.

Kosic, A., & Triandafyllidou, A. (2004). Albanian and Polish migration to Italy: The micro-processes of policy, implementation and immigrant survival strategies. *International Migration Review, 38*(4), 1413–1446.

León-Ledesma, M., & Piracha, M. (2004). International migration and the role of remittances in Eastern Europe. *International Migration, 42*(4), 65–83.

Lindstrom, D. P. (1996). Economic opportunity in Mexico and return migration from the United States. *Demography, 33*(3), 357–374.

Lindstrom, D. P., & Lauster, N. (1999). Local economic opportunity and the competing risks of internal and U.S. migration in Zacatecas, Mexico. *International Migration Review, 35*(4), 1232–1256.

Lippman, S. A., & McCall, J. J. (1976). The economics of job search: A survey. *Part I. Economic Inquiry, 14*, 155–189.

Marques, H. (2010). Migration creation and diversion in the European Union: Is Central and Eastern Europe a 'natural' member of the single market for labour? *Journal of Common Market Studies, 48*(2), 265–291.

Martin, P. (2006). Managing labor migration: temporary worker programmes for the 21st century. International symposium on international migration and development, population division, department of economic and social affairs, United Nations Secretariat, Turin, 28–30 June 2006.

Massey, D. S., Alarcón, R., Durand, J., & González, H. (1987). *Return to Aztlan: The social process of international migration from Western Mexico*. Berkeley: University of California Press.

Massey, D. S., Durand, J., & Malone, N. J. (2002). *Beyond smoke and mirrors: Mexican immigration in an era of economic integration*. New York: Russell Sage Foundation.

Massey, D. S., & Liang, Z. (1989). The long-term consequences of a temporary worker program: The U.S. Bracero experience. *Population Research and Policy Review, 8*, 199–226.

Massey, D. S., & Parrado, E. A. (1998). International migration and business formation in Mexico. *Social Science Quarterly, 79*(1), 1–20.

Mattila, J. P. (1974). Job quitting and frictional unemployment. *The American Economic Review, 64*(1), 235–239.

McCormick, B., & Wahba, J. (2004). Return international migration and geographical inequality: The case of Egypt. United Nations University, World Institute for Development Economics Research, *Research Paper No.* 2004/7.

Montgomery, J. (1991). Social networks and labor-market outcomes: Toward an economic analysis. *American Economic Review, 81*(5), 1408–1418.

Mortensen, D. T. (1986). Job search and labor market analysis. In O. Ashenfelter & R. Layard (Eds.), *Handbook of Labor Economics* (Vol. II, pp. 849–919). New York: Elsevier Science Publisher.

Oucho, J. O. (1998). Recent internal migration processes in Sub-Saharan Africa: Determinants, consequences, and data adequacy issues. In R. E. Bilsborrow (Ed.), *Migration, urbanization, and development: New directions and issues* (pp. 89–120). New York: United Nations Population Fund and Kluwer Academic Publishers.

Papademetriou, D. G., Meissner, D., Rosenblum, M. R., & Sumption, M. (2009). *Aligning temporary immigration visas with US labor market needs: The case for a new system of provisional visas*. Washington: Migration Policy Institute.

Parkes, R. (2009). EU mobility partnerships: A model of policy coordination? *European Journal of Migration and Law, 11*, 327–345.

Parsons, D. O. (1973). Quit rates over time: A search and information approach. *The American Economic Review, 63*(3), 390–401.

Passel, J. S. and Cohn, D. (2009). Mexican immigrants: How many come? How many leave? Washington, DC: Pew Hispanic Center (22 July 2009).

Passel, J. S. and Cohn, D. (2010). U.S. unauthorized immigrant flows are down sharply since mid-decade. Washington: Pew Hispanic Center (1 Sept 2010).

Passel, J. S. & Cohn, D. (2011). Unauthorized immigrant population: National and state trends, 2010. Washington: Pew Hispanic Center (1 Feb 2011).

Peixoto, J. (2009). New migrations in Portugal: Labour markets, smuggling and gender segmentation. *International Migration, 47*(3), 185–210.

Piore, M. J. (1979). *Birds of passage: Migrant labor and industrial societies*. New York: Cambridge University Press.

Pissarides, C. A. (1992). Loss of skill during unemployment and the persistence of employment shocks. *The Quarterly Journal of Economics, 107*(4), 1371–1391.

Plaza, D. (2008). Transnational return migration to the English-speaking Caribbean. *Revue Européenne des Migrations Internationales, 24*(1), 115–137.

Reichert, J. S. (1981). The migrant syndrome: Seasonal U.S. wage labor and rural development in central Mexico. *Human Organization, 40*(1), 56–66.

Reyes, B. I. (2001). Immigrant trip duration: The case of immigrants from Western Mexico. *International Migration Review, 35*(4), 1185–1204.

Ruhs, M., & Anderson, B. (2010). Semi-compliance and illegality in migrant labour markets: An analysis of migrants, employers and the state in the UK. *Population Space and Place, 16*, 195–211.

Ruhs, M., & Martin, P. (2008). Numbers vs. rights: Trade-offs and guest worker programs. *International Migration Review, 42*(1), 249–265.

Rye, J. F., & Andrzejewska, J. (2010). The structural disempowerment of Eastern European migrant farm workers in Norwegian agriculture. *Journal of Rural Studies, 26*, 41–51.

Schwab, S. J. (1999). Employment discrimination. In B. Bouckaert & G. De Geest (Eds.), *Encyclopedia of law and economics* (pp. 1–22). Aldershot: Edward Elgar.

Stark, O. (1991). *The migration of labor.* Cambridge: Basil Blackwell Inc.

Taylor, J. E. (1999). The new economics of labour migration and the role of remittances in the migration process. *International Migration, 37*(1), 63–88.

Ünver, O.C. (2006). Current discussions in the German integration debate: The culturalist vision vs. social equity? *Revue européenne des migrations internationales, 22*(3), 2–14.

U.S. Census Bureau (2009). Current population survey, annual social and economic supplement.

Van Dijk, J., & Folmer, H. (1985). Entry of the unemployed into employment: Theory, methodology and Dutch experience. *Regional Studies, 19*(3), 243–256.

Vasileva, K. (2010). Foreigners living in the EU are diverse and largely younger than the nationals of the EU member states. *Eurostat Statistics in Focus* 45/2010.

Walmsley, T., & Winters, A. L. (2005). Relaxing the restrictions on the temporary movement of natural persons: A simulation analysis. *Journal of Economic Integration, 20*(4), 688–726.

Winters, L. A., Walmsley, T. L., Wang, Z. K., & Grynberg, R. (2003). Liberalising temporary movement of natural persons: An agenda for the development round. *World Economy, 26*(8), 1137–1161.

Wood, C. H. (1981). Structural changes and household strategies: A conceptual framework for the study of rural migration. *Human Organization, 40*(4), 338–344.

Woodruff, C., & Zenteno, R. (2007). Migration networks and microenterprises in Mexico. *Journal of Development Economics, 82*, 509–528.

# Chapter 9
# Event History Analysis: Local Dependence and Cross-Sectional Sampling

**Niels Keiding**

## 9.1 Introduction: Markov Chain Models for Event History Analysis

Jan Hoem's teacher Erling Sverdrup published his important survey on statistical analysis of continuous-time Markov chains for simple mortality and morbidity processes in 1965 (Sverdrup 1965), and Jan followed up in his dissertation with a long series of detailed expositions on how to validate and further develop existing procedures in demography and life insurance mathematics in this framework (e.g. Hoem 1969a, 1971). Much of Jan's substantive demographic research has been based on this *event-history analysis* approach ever since, and I have chosen to present a case study to illustrate some of the methodological and substantive perspectives that still enliven classical event history analysis.

The example comes from an apparently innocent consulting problem presented to us in Copenhagen in 1977, where Jan was professor of actuarial mathematics, his former master student Odd Aalen was visiting on his way back to Norway from his successful Berkeley Ph.D. thesis, and I was a mathematical statistician. (See Aalen et al. (2009) for a description of that fertile scientific environment and some of the results obtained there.) In his out-patient clinic at The Finsen Institute in Copenhagen,[1] Dr. Jens Thormann had collected a consecutive series of patients with the chronic skin disease *pustulosis palmo-plantaris*. Most were women, and he had asked them whether their menopause had already occurred, and if so, when. From an epidemiological point of view this is case-only data with retrospective

[1] The Finsen Institute was a cancer hospital in Copenhagen, which is now merged into the University hospital.

N. Keiding (✉)
Department of Public Health, University of Copenhagen, Copenhagen, Denmark
e-mail: nike@sund.ku.dk

G. Neyer et al. (eds.), *The Demography of Europe*,
DOI: 10.1007/978-90-481-8978-6_9,

information on the possible determinant. We chose to analyze this implementing probabilistic causality ideas of Schweder (1970) ('local dependence') in combination with results by Hoem (1969b) on conditioning in the Markov chains underlying the by then emerging field of event history analysis. The analysis (Aalen et al. 1980) showed clear indication that the age-specific incidence of the disease was larger after menopause than before, and since this result was new and unexpected in the dermatological literature, we arranged for a confirmatory data set to be collected at a different clinic. In this chapter I present the latter results, which indicate that the finding may stand. Jan discussed our analysis with us and has since then often emphasized the desirability of developing a systematic approach to handling sampling patterns in event history analysis. In Sect. 9.2 I outline the main features of an approach to a subset of such problems: sampling a developing population at a cross-section in calendar time, using the Lexis diagram as the indispensable tool (Keiding 2006), deferring the case study proper to Sect. 9.3. A newer, alternative approach to case-only data is the so-called case series analysis (e.g. Farrington and Whitaker 2006), which may also be illustrated in the present study, see Sect. 9.4. Section 9.5 is a brief conclusion.

## 9.2 Sampling Event Histories at a Cross-Section

In an event-history study of diseases, a population of individuals is modelled by starting an illness-death process (Fig. 9.1) at each individual's calendar time of birth and following it through to possible disease and ultimate death. Figure 9.2 is a Lexis diagram representation of this model, with the additional feature of sampling the population at calendar time $t_0$, a so-called cross-section.

Keiding (1991, 2006), see also Lund (2000), surveyed a class of problems in observational epidemiology that could all be viewed as special cases of this type of sampling, including current status data, retrospective incidence estimation, current duration data and prevalent cohort sampling.

*Current status data* record the ages at $t_0$ of the healthy and of the diseased. Under the restrictive assumptions of no differential mortality between healthy and diseased and no calendar time effects on birth, incidence and mortality it is possible to estimate the incidence rate $\alpha(a)$, in practice the corresponding distribution function $1 - \exp\left(-\int_0^a \alpha(u)du\right)$.

*Retrospective incidence estimation* can sometimes be based on information about onset ages for each individual in the prevalent sample. Since individuals are included conditional on having survived until the sampling date $t_0$, information on

**Fig. 9.1** The basic illness-death process

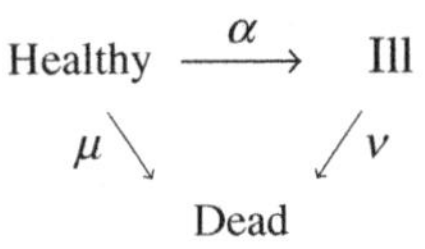

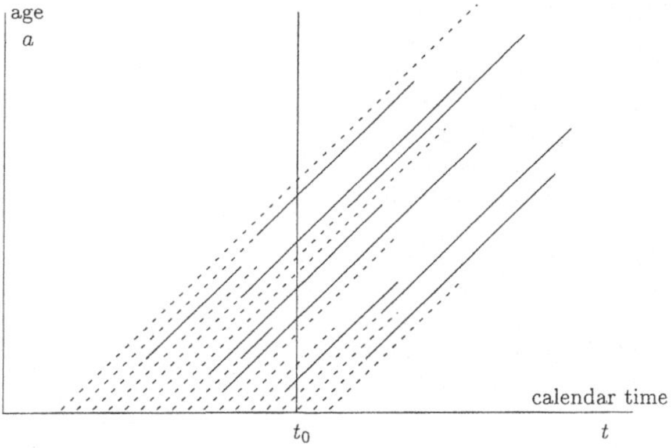


**Fig. 9.2** Lexis diagram of individuals born healthy (*dotted line*), possibly becoming ill (*solid line*) and dying, when a cross-section is taken at time $t_0$

mortality during the years before $t_0$ is required so that each individual may be weighted [in a Horvitz and Thompson (1952) fashion] with the inverse probability of being sampled. Retrospective information (e.g. census data) on the population at risk is also necessary. Keiding et al. (1989) presented an early case study concerning historical incidence of diabetes. Such calculations are still being developed, see e.g. Alioum et al. (2005) and Addona et al. (2009).

*Current duration data* or backward recurrence times are used to estimate mortality from the age distribution of the living at the cross-section, assuming stationarity. The basic idea is well known in demography: in a stationary population the life table is proportional to the age distribution of the living. See e.g. Hald (1989) for an exposition of the use of this concept by Halley (1693). More recently, such data are termed "open-interval" or "last-episode" cf. Yamaguchi (2003), who was inspired by migration analysis. Keiding et al. (2002) and Scheike and Keiding (2006) surveyed application of this approach to time-to-pregnancy studies, and Slama et al. (2006) gave a case study.

Mortality estimation from follow-up of a *prevalent cohort* is still an active research area in biostatistics, even if the technique to a demographer is an obvious version of classical life table calculations.

Let me briefly recall the concepts of *left truncation* and *delayed entry* in survival analysis. The basic problem in survival analysis is to estimate the distribution of a positive random variable $X$, the *survival time*, with density $f(x)$, survival function $S(x) = \int_x^\infty f(u)du$ and hazard rate $\varphi(x) = f(x)/S(x)$. Under left truncation at $v$, we do not observe $X$ but the conditional distribution of $X$ given $X > v$. However, for $X > v$ the hazard function of this conditional distribution is seen to be the same as that of $X$, so hazard-based survival analysis methods also apply to left truncated data, as long as the risk sets are adjusted accordingly, just as in ordinary life table calculations. The fundamental paper by Kaplan and Meier (1958) acknowledged this fact, but many later authors (e.g. Tsai et al. 1987) considered it a new finding.

Returning to the prevalent cohort study, we assume that the target is the mortality of the diseased or *lethality* $v(t, a, d)$ (cf. Fig. 9.1) which in general

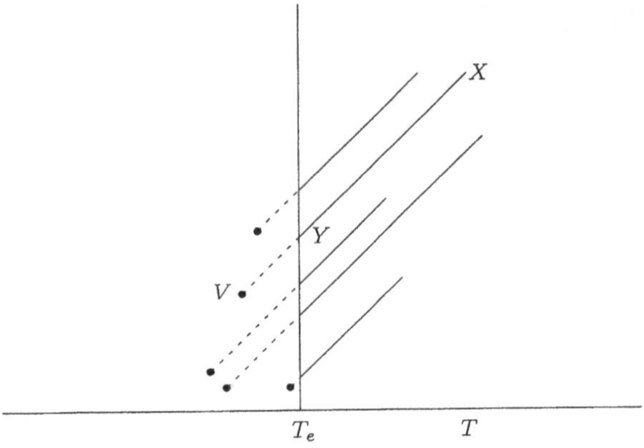


**Fig. 9.3** Lexis diagram illustrating follow-up of the diseased, with: $V$ age at onset of disease (assumed known), *dotted line* non-observed time between age at onset of disease and age at entry into survey, $Y$ age at entry into survey, *solid line* observed time between age at entry into survey and age at time of death, $X$ age at death, $T_e$ time at entry, $T$ time at death

depends on calendar time $t$, age $a$ and disease duration $d$. A standard version of the problem based on the Lexis diagram in Fig. 9.2 is outlined in Fig. 9.3. Let $V =$ age at disease onset, $Y =$ age at entry into the study (i.e. at time $t_0$), $X =$ age at death, $T =$ time at death; then $T_e = T - (X - V)$ is time at entry ($t_0$ above). It may then be shown by a simple concrete calculation (Keiding 1992) that the intensity that a diseased individual dies at time $t$ and age $x$, given disease onset at age $v < x$ and entry into study at age $y > v$ equals the lethality $\nu(t, x, x - v)$ for $x > y$. In other words, despite the length-biased recruitment of prevalent patients into the prevalent cohort, standard survival analysis with delayed entry applies.

It should be mentioned that the above exposition works conditional on the realised prevalent sample. Since individuals with long disease durations are overrepresented in the prevalent sample ('length bias'), there might be further information in including the distribution of the sample in the analysis. See Wang (1991) and Bergeron et al. (2008) and their references.

Wang et al. (1993), cf. Ripley and Solomon (1995) discussed the use of Cox regression models for prevalent cohort data. For these models the choice of time origin for the time variable entering into the underlying intensity is important and the use of time-dependent covariates requires special care (Keiding and Knuiman 1990). Andersen et al. (1985) presented a careful case study on survival of a prevalent cohort of diabetics using a special version of Cox regression modelling of the *relative* mortality.

The above set-up about statistical inference of survival data based on sampling on the Lexis diagram has recently been generalised by Guilloux (2007) and Brunel et al. (2008).

In sum, event history analysis has always been characterized by the impossibility of following up every individual and the resulting ubiquitous occurrence of *censoring*. As often pointed out by Jan Hoem, too little emphasis has been put on adjusting the methods of analysis to the additional problem of *non-standard sampling patterns*, in demography and elsewhere. The above outlines some recent approaches to solutions to this challenge.

## 9.3 Retrospective Analysis of the Interaction of Life History Events

The problem outlined in the introduction of assessing the possible effect of menopause on the incidence of *pustulosis palmo-plantaris* from a sample of the prevalent population of patients with this disease may be formalised as shown in Fig. 9.4.

To transfer that into a tractable event history analysis framework requires several steps. Aalen et al. (1980) started from the important concept due to Schweder (1970) of *local dependence* in terms of *composable Markov processes*: assume that transitions to states $A$ and $B$ as well as the order of these transitions are represented by a four-state Markov process as in Fig. 9.5.

If $\alpha_{0B} \equiv \alpha_{A,AB}$ but $\alpha_{0B} < \alpha_{B,AB}$ we say that $A$ is locally dependent on $B$ but $B$ is not locally dependent on $A$.

Aalen et al. (1980) considered 85 consecutive female patients with *pustulosis palmo-plantaris.* For each patient the age of onset ($A$) and age of menopause ($B$) (if it had yet happened) were registered. To assess whether local dependence in the above sense could be assessed from these retrospective "case-only" data, the model in Fig. 9.5 was elaborated with "sampled" and "death" states as in Fig. 9.6 and the conditional process, given termination in $SA$ or $SAB$ (or equivalently, in $\dagger_S$), was considered.

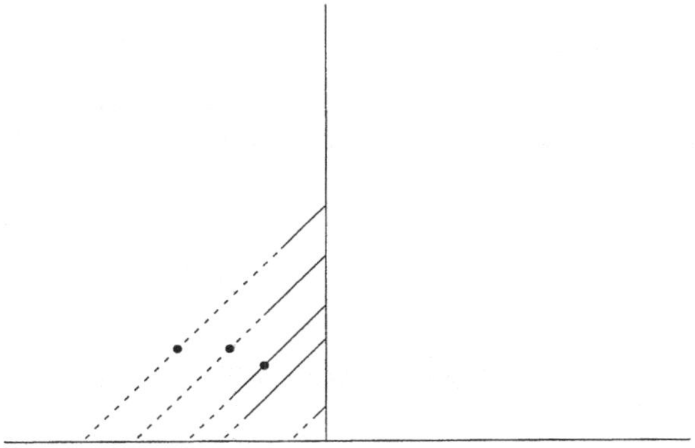

**Fig. 9.4** Lexis diagram illustrating retrospective estimation of interaction between disease incidence and another life history event, here menopause (*filled circle*). Notation as in Fig. 9.2, i.e. healthy (*dotted line*) and ill (*solid line*)

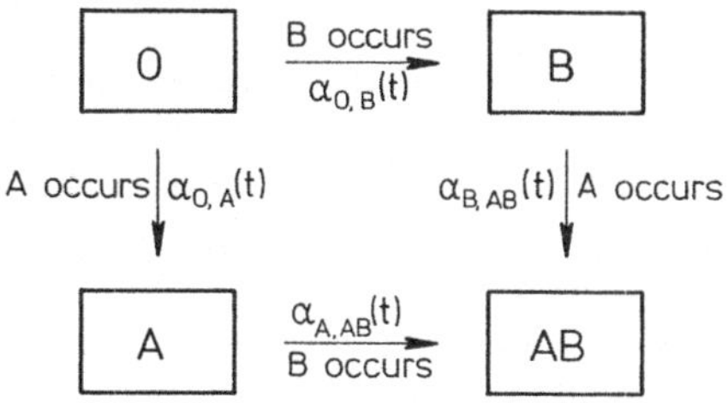


**Fig. 9.5** Markov process model for the interaction of two life history events [reproduced from Aalen et al. (1980) with the kind permission of Blackwell Publishing]

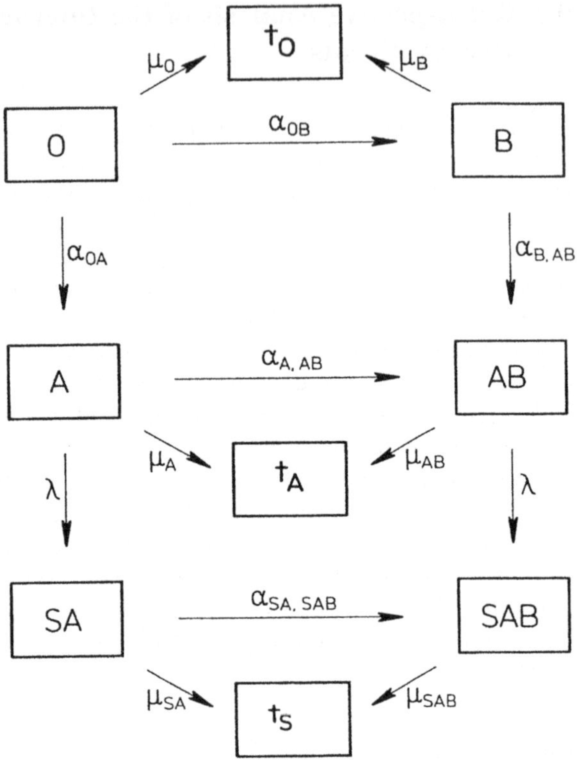


**Fig. 9.6** The Markov model of Fig. 9.5, extended with "sampled" and "dead" states to model cross-sectional sampling of the population of diseased. *S* sampled, † dead [reproduced from Aalen et al. (1980) with the kind permission of Blackwell Publishing]

As shown in the important but little-known note by Hoem (1969b), this is again a Markov process with new transition intensities

$$\gamma_{ij}(t) = \alpha_{ij}(t)\frac{\Pi_j(t)}{\Pi_i(t)}$$

where

$$\Pi_k(t) = P_{k\dagger_s}(t, \infty) = P(\text{patient in state } k \text{ at age } t \text{ will eventually get sampled}).$$

Aalen et al. (1980) assumed non-differential mortality

$$\mu_0 \equiv \mu_A \equiv \mu_B \equiv \mu_{AB}$$

and gave a detailed discussion on conditions under which

$$\alpha_{0A} \equiv \alpha_{B,AB} \Rightarrow \gamma_{0A} \equiv \gamma_{B,AB}$$
$$\alpha_{0B} \equiv \alpha_{A,AB} \Rightarrow \gamma_{0B} \equiv \gamma_{A,AB}$$

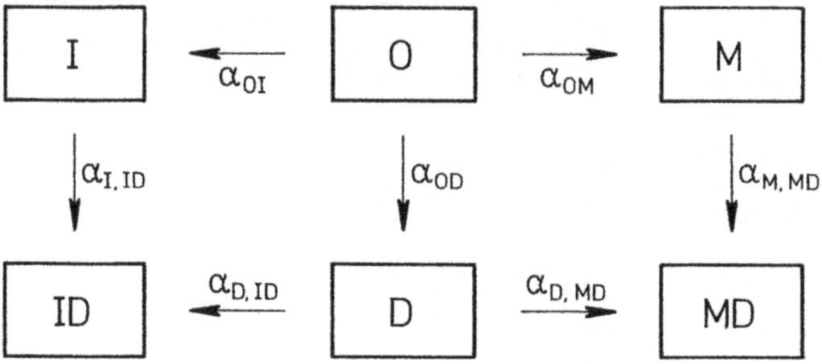


**Fig. 9.7** Markov chain model for the occurrence of natural and induced menopause, and the outbreak of pustulosis palmo-plantaris. *0* no event has occurred; *M* (natural) menopause has occurred; *D* disease has broken out; *I* induced menopause has occurred [reproduced from Aalen et al. (1980) with the kind permission of Blackwell Publishing]

so that (conservative) tests of the hypotheses $\alpha_{0A} \equiv \alpha_{B,AB}$ and $\alpha_{0\mathrm{B}} \equiv \alpha_{A,AB}$ may be performed by standard hypothesis tests in the Markov process obtained from the prevalent sample.

In the concrete example this procedure was developed for the more complicated model distinguishing between natural and induced menopause, cf. Fig. 9.7.

The result was a clear rejection ($P < 0.0005$) of the hypothesis

$$\alpha_{0D} \equiv \alpha_{I,ID} \equiv \alpha_{M,MD}$$

indicating that natural as well as induced menopause seem to increase the incidence of pustulosis palmo-plantaris. Supported by close similarity of the observed menopause rates to data from the general population, there was also judged to be strong indication that $\alpha_{0M} = \alpha_{D,MD}$ and $\alpha_{0I} = \alpha_{D,ID}$ (disease occurrence does not influence occurrence of menopause).

A second sample of consecutive patients with *pustulosis palmo-plantaris*, this time with explicit exclusion of patients suspected for psoriasis, was collected by Dr. Marie Cramers at the Marselisborg Hospital in- and out-patient clinics in Århus, Denmark, over the period 1979, 1980, and the early months of 1981. There were 11 men and 70 women, the latter having first appearance, menopause and last seen as shown in Table 9.1, following the same layout as Aalen et al. (1980, Table 1). The observed flows in the six-state event history models are quite similar for the two datasets (Fig. 9.8), except that there were rather fewer induced menopauses in Marselisborg. The estimates of the cumulative intensities $\int_0^t \gamma_{ij}(u)du$ also follow the same patterns as for the original (Finsen) data (not shown here). The conservative three-sample test of the hypothesis $\alpha_{0D} \equiv \alpha_{I,ID} \equiv \alpha_{M,MD}$ this time yields $P = 0.032$.

In previous research, Courgeau and Lelièvre (1992, Chap. 5) carefully described the importance in demography of being able to study the interaction between two events using marriage and migration as their example. Blossfeld and Rohwer (2002) followed suit. Recently, Hoem and Nedoluzhko (2008) gave an updated discussion

**Table 9.1** Age of first appearance, age of menopause or induced (*) menopause and age last seen for 70 women with pustulosis palmo-plantaris seen at Marselisborg Hospital

| First appear | Menopause | Last seen |
|---|---|---|
| 14 | – | 16 |
| 20 | – | 30 |
| 21 | – | 22 |
| 21 | – | 27 |
| 21 | – | 50 |
| 22 | – | 27 |
| 24 | – | 28 |
| 25 | – | 43 |
| 32 | – | 36 |
| 33 | 52 | 62 |
| 35 | – | 41 |
| 35 | – | 45 |
| 35 | 36* | 69 |
| 37 | – | 37 |
| 40 | – | 47 |
| 40 | 45 | 60 |
| 42 | 37 | 45 |
| 42 | 40 | 58 |
| 43 | 33 | 58 |
| 44 | – | 50 |
| 45 | – | 47 |
| 45 | – | 48 |
| 45 | 49 | 53 |
| 46 | 45 | 48 |
| 46 | 52 | 70 |
| 47 | 46 | 48 |
| 47 | 44 | 52 |
| 47 | 49 | 54 |
| 48 | – | 52 |
| 48 | 40 | 63 |
| 49 | 52 | 52 |
| 49 | – | 51 |
| 50 | 49 | 51 |
| 50 | 49 | 52 |
| 50 | 49 | 60 |
| 51 | – | 51 |
| 51 | 30* | 53 |
| 51 | 45* | 54 |
| 51 | 48 | 56 |
| 51 | 50 | 64 |
| 52 | 47 | 54 |
| 52 | 43 | 57 |
| 52 | 51 | 64 |
| 53 | 43 | 53 |

(continued)

**Table 9.1** (continued)

| First appear | Menopause | Last seen |
|---|---|---|
| 53 | 46* | 54 |
| 54 | 53 | 56 |
| 54 | 50 | 65 |
| 55 | 36 | 62 |
| 55 | 50* | 59 |
| 56 | 37 | 57 |
| 56 | 53 | 57 |
| 56 | 49 | 58 |
| 56 | 50 | 69 |
| 57 | 48 | 58 |
| 57 | 51 | 67 |
| 58 | 48* | 59 |
| 58 | 46 | 61 |
| 59 | 56 | 60 |
| 59 | 49 | 79 |
| 60 | 44 | 62 |
| 61 | 51 | 63 |
| 63 | 41 | 66 |
| 63 | 49 | 71 |
| 63 | 50 | 63 |
| 65 | 59 | 70 |
| 70 | 45 | 71 |
| 70 | 51* | 72 |
| 70 | 35* | 79 |
| 73 | 50 | 74 |
| 74 | 55 | 78 |

of marriage and migration, using the modelling presented here. To conclude, there are two issues to consider: (1) the conceptual modelling of the interaction pattern, going back to Schweder (1970) and being very similar in spirit to that of Granger causality, and (2) the actual statistical analysis of these models, taking into account the precise observational pattern. Once the modelling is in place, most of the analysis may be performed by means of standard event history analysis allowing for censoring (generalized survival analysis), now a standard tool in demography. However, possible non-standard sampling patterns may require modifications, such as reviewed above and illustrated by Aalen et al. (1980).

## 9.4 Case Series Analysis Approach

In this section, I outline an alternative approach to the analysis of *'case-only data'* such as those of the skin disease patients analysed by Aalen et al. (1980).

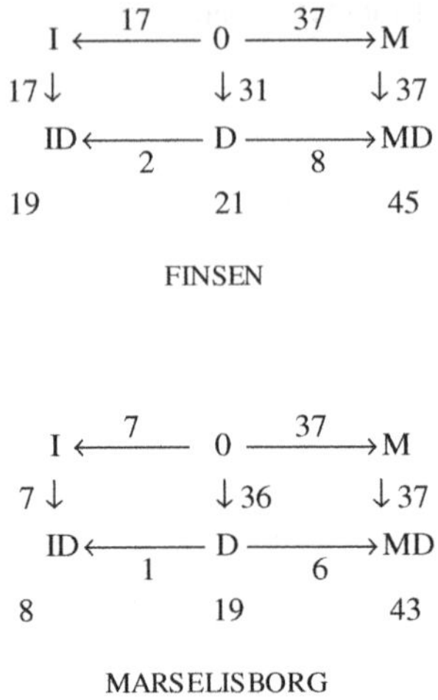


**Fig. 9.8** Transitions between states and numbers in ID, D and MD at the survey

Farrington and Whitaker (2006) proposed a proportional hazards model allowing direct estimation of the relative rates

$$\beta_M = \alpha_{M,MD}(a)/\alpha_{0D}(a) \text{ and } \beta_I = \alpha_{I,ID}(a)/\alpha_{0D}(a)$$

of disease incidence at age $a$ after (natural or induced) menopause compared to that before.

The observation scheme may be defined from three latent ages

$a_I$ age at induced menopause
$a_M$ age at natural menopause
$a_C$ age last seen

We only know $a_I$ if $a_I < a_M \wedge a_C$, we only know $a_M$ if $a_M < a_I \wedge a_C$, we always know $a_C$, and we only observe the woman if disease incidence occurred before $a_C$. Following Farrington and Whitaker, we assume a semiparametric model of disease incidence given the individual history $x_i$ which contains menopause career $\left(a_I^i, a_M^i\right)$ and censoring pattern $\left(a_C^i\right)$ for individual $i$ and which is assumed completely independent (exogenous) of the disease event. This model is given by the disease incidence at age $a$

$$\begin{aligned} \lambda_i(a|x_i) = \psi(a), & \quad 0<a<a_M^i \wedge a_I^i \wedge a_C^i \\ \beta_I\psi(a),_I^i & <a<a_C^i,\ a_I^i<a_M^i \\ \beta_M\psi(a),_M^i & <a<a_C^i,\ a_M^i<a_I^i \end{aligned}$$

We assume the disease *rare* in the technical sense that

$$\lambda(a|x_i) = \varphi v(a|x_i)$$

where $\varphi$ is small and $v$ is bounded in $a$. Defining the survival function

$$S(a|x_i) = \exp\left(-\int_0^a \lambda(t|x_i)dt\right)$$

we have, as $\varphi \to 0$, that

$$S(a|x_i) \to 1$$

$$S(0|x_i) - S(a|x_i) \approx \int_0^a \lambda(t|x_i)dt.$$

The conditional likelihood of patient $i$ expresses the conditional probability of disease having happened at age $a_i$ given menopause and censoring history $x_i$ and that disease happened in $\left(0, a_C^i\right)$, i.e.

$$L_i = \frac{S(a_i|x_i)\lambda(a_i|x_i)}{S(0|x_i) - S\left(a_C^i|x_i\right)} \approx \frac{\lambda(a_i|x_i)}{\int_0^{a_C^i} \lambda(a|x_i)da}$$

The product of these for all patients $i$ yields a partial likelihood in $\beta_I$ and $\beta_M$. Using the software provided by Whitaker et al. (2006) we obtain estimates with 95 % confidence intervals of the relative disease incidence rates:

| | Finsen study | Marselisborg study |
|---|---|---|
| Natural menopause | | |
| $\hat{\beta}_M$ | 4.2 (1.2, 14.9) | 3.4 (1.0, 11.2) |
| Induced menopause | | |
| $\hat{\beta}_I$ | 14.0 (1.9, 104.4) | 1.6 (0.2, 17.3) |

An approximate Hotelling's $T^2$-test for identity of the original (Finsen) and confirmatory (Marselisborg) study yields $P = 0.4$.

It is certainly comforting that the results of this re-analysis agree so well with the principal results from the event history analysis presented in Sect. 9.3.

The case-series analysis approach has existed for some time but has recently been considerably refined and generalized by C. P. Farrington and colleagues. Its main applications have so far been in epidemiology, in particular in vaccination studies and in pharmacoepidemiology. I included this application to the skin disease example in order to connect to the above presentation, but also, more generally, to illustrate the additional possibilities it offers: the definition and estimation of an easily interpretable effect parameter and the easy accommodation of covariates. It remains to be seen how important the case series approach will become in demographic research.

## 9.5 Conclusions

This brief contribution has attempted to show that there is still considerable methodological activity in two aspects of event history analysis identified by Jan Hoem many years ago as important: statistical inference under non-standard sampling patterns, and the local dependence concept akin to that of Granger causality (Granger 1969).

The local dependence concept was formulated in 1970 by Schweder in the context of Markov processes. Aalen (1987), cf. Mykland (1986) in an unpublished technical report, generalised the concept beyond the original Markov process framework; but Aalen's definition still only considers two processes at a time. Multivariate generalisations were described by Courgeau and Lelièvre (1992, Chap. 6) and in a more general graph-theoretic framework by Didelez in her dissertation (Didelez 2000), from which the main results have recently appeared in the journal literature (Didelez 2007, 2008).

The main message to demographers of this contribution is that the models describing interaction patterns as those demonstrated here are very natural and obvious generalizations of basic event history analysis. As long as standard sampling patterns have generated the data for analysis, the analyses are obvious generalizations of well-known techniques. However, sometimes data have been sampled in more complicated ways, and then more intricate analyses are necessary.

**Acknowledgments** Our thanks are due to Marie Cramers for collecting the confirmatory data on *pustulosis palmo-plantaris* and to Ørnulf Borgan and Randi Grøn for help with computations. This research was partially supported by a grant (RO1CA54706-12) from the National Cancer Institute and by the Danish Natural Sciences Council grant 272-06-0442 "Point process modelling and statistical inference".

## References

Aalen, O. O. (1987). Dynamic modelling and causality. *Scandinavian Actuarial Journal, 12*, 177–190.

Aalen, O. O., Borgan, Ø., Keiding, N., & Thormann, J. (1980). Interaction between life history events: Nonparametric analysis of prospective and retrospective data in the presence of censoring. *Scandinavian Journal of Statistics, 7*(4), 161–171.

Aalen, O. O., Andersen, P. K., Borgan, Ø., Gill, R., & Keiding, N. (2009). History of applications of martingales in survival analysis. *Electronic Journal for History of Probability and Statistics, 5*(1), 1–28.

Addona, V., Asgharian, M., & Wolfson, D. B. (2009). On the incidence-prevalence relation and length-biased sampling. *Canadian Journal of Statistics, 37*(2), 206–218.

Alioum, A., Commenges, D., Theibaut, R., & Dabis, F. (2005). A multistate approach for estimating the incidence of human immunodeficiency virus by using data from a prevalent cohort study. *Journal of the Royal Statistical Society Series C, 54*(4), 739–752.

Andersen, P. K., Borch-Johnsen, K., Deckert, T., Green, A., Hougaard, P., Keiding, N., et al. (1985). A Cox regression model for the relative mortality and its application to diabetes mellitus survival data. *Biometrics, 41*(4), 921–932.

Bergeron, P.-J., Asgharian, M., & Wolfson, D. B. (2008). Covariate bias induced by length-biased sampling of failure times. *Journal of the American Statistical Association, 103*(482), 737–742.

Blossfeld, H.-P., & Rohwer, G. (2002). *Techniques of event history modeling*. Mahwah: Erlbaum.

Brunel, E., Comte, F., & Guilloux, A. (2008). Estimation strategies for censored lifetimes with a Lexis-diagram type model. *Scandinavian Journal of Statistics, 35*(3), 557–576.

Courgeau, D., & Lelièvre, É. (1992). *Event history analysis in demography*. Oxford: Clarendon Press.

Didelez, V. (2000). *Graphical models for event history data based on local independence*. Berlin: Logos.

Didelez, V. (2007). Graphical models for composable finite Markov processes. *Scandinavian Journal of Statistics, 34*(1), 169–185.

Didelez, V. (2008). Graphical models for marked point processes based on local independence. *Journal of the Royal Statistical Society B, 70*(1), 245–264.

Farrington, C. P., & Whitaker, H. J. (2006). Semiparametric analysis of case series data. *Journal of the Royal Statistical Society Series C, 55*(5), 553–580.

Granger, C. W. J. (1969). Investigating causal relations by econometric models and cross-spectral methods. *Econometrica, 37*(3), 424–438.

Guilloux, A. (2007). Nonparametric estimation for censored lifetimes suffering from an unknown selection bias. *Mathematical Methods in Statistics, 16*(3), 202–216.

Hald, A. (1989). *A history of probability and statistics and their applications before 1750*. New York: Wiley.

Halley, E. (1693). An estimate of the degrees of the mortality of mankind drawn from curious tables of the births and funerals of the city of Breslaw. *Philosophical Transactions of the Royal Society of London, 17*, 596–610.

Hoem, J. M. (1969a). Fertility rates and reproduction rates in a probabilistic setting. *Biométrie-Praximétrie, 10*(1), 38–66. Correction vol 11 (1970): 20.

Hoem, J. M. (1969b). Purged and partial Markov chains. *Skandinavisk Aktuarietidskrift, 52*(45), 147–155.

Hoem, J. M. (1971). Point estimation of forces of transition in demographic models. *Journal of the Royal Statistical Society B, 33*(2), 275–289.

Hoem, J. M., & Nedoluzhko, L. (2008). Marriage formation as a process intermediary between migration and childbearing. *Demographic Research, 18*(21), 611–628.

Horvitz, D. G., & Thompson, D. J. (1952). A generalization of sampling without replacement from a finite universe. *Journal of the American Statistical Association, 47*(260), 663–685.

Kaplan, E. L., & Meier, P. (1958). Non-parametric estimation from incomplete observations. *Journal of the American Statistical Association, 53*(282), 457–481, 562–563.

Keiding, N. (1991). Age-specific incidence and prevalence: A statistical perspective (with discussion). *Journal of the Royal Statistical Society A, 154*(3), 371–412.

Keiding, N. (1992). Independent delayed entry (with discussion). In: J. P. Klein & P. K. Goel (Eds.), *Survival analysis: State of the art* (pp. 309–326). Dordrecht: Kluwer.

Keiding, N. (2006). Event history analysis and the cross-section. *Statistics in Medicine, 25*(14), 2343–2364.

Keiding, N., & Knuiman, M. W. (1990). Letter to the editor on "survival analysis in natural history studies of disease" by Cnaan, A. and Ryan, L. *Statistics in Medicine, 9*(10), 1221–1222.

Keiding, N., Holst, C., & Green, A. (1989). Retrospective estimation of diabetes incidence from information in a current prevalent population and historical mortality. *American Journal of Epidemiology, 130*(3), 588–600.

Keiding, N., Kvist, K., Hartvig, H., Tvede, M., & Juul, S. (2002). Estimating time to pregnancy from current durations in a cross-sectional sample. *Biostatistics, 3*(4), 565–578.

Lund, J. (2000). Sampling bias in population studies—How to use the Lexis diagram. *Scandinavian Journal of Statistics, 27*(4), 589–604.

Mykland, P. (1986). Statistical causality. *Technical Report 14*. Department of Mathematics, University of Bergen, Norway.

Ripley, B. D., & Solomon, P. J. (1995). Correspondence: Statistical methods for prevalent cohort data. *Biometrics, 51*(1), 373–375.

Scheike, T. H., & Keiding, N. (2006). Design and analysis of time to pregnancy. *Statistical Methods in Medical Research, 15*(2), 127–140.

Schweder, T. (1970). Composable Markov processes. *Journal of Applied Probability, 7*(2), 400–410.

Slama, R., Ducot, B., Carstensen, L., Lorente, C., de La Rochebrochard, E., Leridon, H., et al. (2006). Feasibility of the current duration approach to study human fecundity. *Epidemiology, 17*(4), 440–449.

Sverdrup, E. (1965). Estimates and test procedures in connection with stochastic models for deaths, recoveries and transfers between different states of health. *Skandinavisk Aktuarietdskrift, 48*(3–4), 184–211.

Tsai, W.-Y., Jewell, N. P., & Wang, M.-C. (1987). A note on the product-limit estimator under right censoring and left truncation. *Biometrika, 74*(4), 883–886.

Wang, M.-C. (1991). Nonparametric estimation from cross-sectional survival data. *Journal of the American Statistical Association, 86*(413), 130–143.

Wang, M.-C., Brookmeyer, R., & Jewell, N. P. (1993). Statistical models for prevalent cohort data. *Biometrics, 49*(1), 1–11.

Whitaker, H. J., Farrington, C. P., Spiessens, B., & Musonda, P. (2006). Tutorial in biostatistics: The self-controlled case series method. *Statistics in Medicine, 25*(10), 1768–1797.

Yamaguchi, K. (2003). Accelerated failure-time mover-stayer regression models for the analysis of last episode data. *Sociological Methodology, 33*(1), 81–110.

# Index




G. Neyer et al. (eds.), *The Demography of Europe*,
DOI: 10.1007/978-90-481-8978-6,

The manufacturer's authorised representative in the EU is Springer Nature Customer Service Centre GmbH, Europaplatz 3, 69115 Heidelberg, Germany. If you have any concerns regarding our products, please contact ProductSafety@springernature.com

Printed and bound by CPI Group (UK) Ltd, Croydon, CR0 4YY
30/09/2026
02225444-0010